ALGEBRA EXAMPLES

POWERS AND LOGARITHMS 2

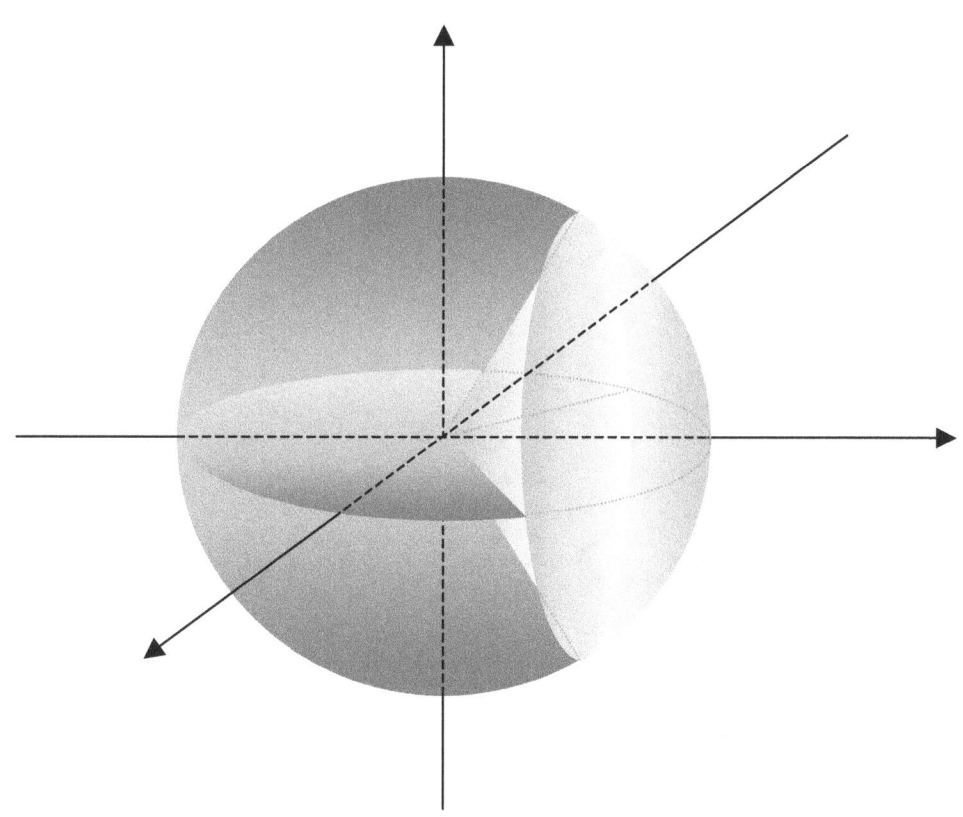

Seong R. KIM

Dear students:

Students need the best teacher, so you need examples, because examples are the best teacher. All the examples here are fully worked, and explain **how** the basic and essential tools in math are made, together with **what** they are, **how** they work, and **how** to work with them. Such tools include numbers, formulas, identities, equations, laws, etc.

Examples here begin with easy ones, of course. Covering every meter and yard properly, we can cover thousands of miles and kilometers. And it is particularly the case in math.

Of those examples therefore, some might even look too easy for you. It's not that easy though, to come up with those examples. Anyways, the bigger and the taller the tree, the deeper and the stronger the root.

Doing math, we work with ideas and run ideas, because every thing in math is an idea. A number is an idea, for instance, and the same is true for a line or circle, too. And putting ideas together, we build another, which becomes the base or an element of another, and each is connected. And that's the way your math grows. So you get to build a circuit, and sometimes, need to fill the gap or repair the circuit so that you get the sense of it.

So your calculation runs properly, and you get the problem solved.

The examples have been made and arranged so that they get tougher (or sometimes easier for some reason) as you proceed with them. In particular, similar examples with some variations are strategically repeated so that you can get the ideas or the tools tricky or complicated, and can get them mastered.

This book is however, nothing but a bunch of examples until you get it powered.   How then, to get it powered, and make it run and work for you?

Just read it, and then, do each example in writing. And it is important to note that you do it in **your** writing. Just watching someone doing it, you just only feel that you can do it. If you do it, you can do it, but if you don't, we can hardly. It's a cliché, of course, but is always true that knowing is one thing and doing is another.

I've been helping students grow, take care of, and run their own math. The area covers algebra and geometry for high school or college students, and is especially for equations (for unknowns or curves), functions, and their graphs, which are the basic elements in calculus, which's been the core of my interest from my early age in high school.

Of my students, some are quite poor in math, and thus, are afraid of or hate math, some require special education because of exceptional intelligence, some are smart enough, some are naïve and diligent, some are clever but lazy, and most behave in general. All the students are badly after though, one thing in common: a strong and secure math skill. It is of course, the prime objective of my work, and I'm always happy to and eager to help them achieve it. The problem was however, that many of them wanted it to be purchased. And the question is, can we buy it?

We can buy the means, of course. And a solid math skill is feasible, too. We know however, we can't buy love, and the same is true for the math skill, too. It's not what we can buy or sell, and not what we can give or take. It is however, what we can grow, and need to grow. Your math grows as much as you grow and take care of it. So does mine.

What math then, do students most often do or use in high schools or colleges?

It is algebra and geometry. What algebra though?

Elementary algebra, of course
Doing the algebra, we work with numbers (many in kinds), constants, variables, ratios, rates, expressions, equations, inequalities, functions, identities, formulas, laws, etc., together with signs and symbols. And if we want to do algebra properly, we want to know their natures and how they mingle with each other.

So studying math ideas or tools, you want to know **what** they are, **how** they work, and **how** to work with them or **what** to do with them.   What then, about the geometry?

Basically, the geometry has much to do with shapes, positions, and angles. The shapes begin with triangles and circles, and move on to rectangles, squares, parallelograms or rhombuses, trapezoids, tetragons, other polygons, polyhedrons, etc.

Doing the geometry, too, though, we need to do the algebra stated above. So it is analytic geometry, often called coordinate geometry, too. And doing it, we can specify positions using coordinates. So in the geometry, basically, we work with graphs. Putting a math idea in a graph, we can not only effectively think about it but actually see it, too, and therefore, can efficiently work with it. What idea then, is it?

The idea begins with a point, line, parabola, circle, ellipse, and hyperbola, called a conic section or basic curve, and then, moves on to other curves, planes, surfaces, volumes, and other objects in various dimensional spaces, together with vectors.

And using an angle, we can specify an amount of turn or change in direction.

So learning, using, or applying those ideas or math tools, we get to solve problems.

And this book can help. It can help learn them, and use them so that you can navigate to find solutions to problems. And in particular, it can help come up with answers to those **what**s and **how**s stated above. So it can help you grow and run your own math, and thus, can help achieve your solid math skill.

It is however, not a magic book giving you a math skill of high caliber overnight. And it can have many mistakes, too. There is no magic, and math is full of facts and ideas. And it is after all, not me and not your teacher but you who put together some of those facts and ideas, and understand it. Putting facts and ideas together, understanding it, and taking care of what you have learned, you grow your math. And this book can help.

This is a book of examples designed to help you grow your math, and assumes that you are a real beginner. This book requires though, time and effort, the amount of which need to be substantial, too, but will be worth it. That's because you want a substantial achievement, and will get it. And probably, you will get to see this book helping you get there much faster than expected. And then, you will get to see the way math runs.

In math, everything is an idea. So is a problem. And solving it, we put it many different ways. For instance, while expanding or reducing it, or modifying or converting it, we keep searching for the solution, approaching the solution, and eventually, can get there. So don't look for the solution outside the problem. The solution is inside the problem if the problem is properly made.

If it is not, no solution is the solution. And in fact, it is often the case a problem itself is the solution. We can put a problem in many different ways, and eventually, can end up with the solution.　　How come then, is the solution no other than the problem?

For instance, the solution to $3232 \div 101$ is 32.　　And we can put it this way:

$$3232 \div 101 = \frac{3232}{101} = \frac{32 \times 101}{101} = \frac{32}{1} = 32 \;\Rightarrow\; 3232 \div 101 = 32.$$

And we can get this, too: $32 \Rightarrow 3232 \div 101$.　　How?

$$32 = \frac{32}{1} = \frac{32 \times 101}{101} = \frac{3232}{101} = 3232/101 = 3232 \div 101.　　\text{Too easy?}$$

For another instance, the solution to $ax^2 + bx + c = 0$ is: $x = \frac{-b \pm \sqrt{b^2 - 4ac}}{2a}$, which is called the quadratic formula.　　How come then, is the solution no other than the problem?

We can put it this way:

$$x = \frac{-b \pm \sqrt{b^2 - 4ac}}{2a} \Rightarrow 2ax = -b \pm \sqrt{b^2 - 4ac} \Rightarrow 2ax + b = \pm\sqrt{b^2 - 4ac}$$

$$\Rightarrow (2ax + b)^2 = b^2 - 4ac \Rightarrow 4a^2x^2 + 4abx + b^2 = b^2 - 4ac$$

$$\Rightarrow 4a^2x^2 + 4abx = -4ac \Rightarrow ax^2 + bx = -c \Rightarrow ax^2 + bx + c = 0.$$

And we can get this, too: $ax^2 + bx + c = 0 \Rightarrow x = \frac{-b \pm \sqrt{b^2 - 4ac}}{2a}$.　　How?

$$ax^2 + bx + c = a(x^2 + \tfrac{b}{a}x) + c = a(x^2 + \tfrac{b}{a}x + \tfrac{b^2}{4a^2} - \tfrac{b^2}{4a^2}) + c = a(x^2 + \tfrac{b}{a}x + \tfrac{b^2}{4a^2}) - \tfrac{b^2}{4a} + c$$

$$= a(x + \tfrac{b}{2a})^2 - \tfrac{b^2 - 4ac}{4a} = 0 \Rightarrow a(x + \tfrac{b}{2a})^2 = \tfrac{b^2 - 4ac}{4a} \Rightarrow (x + \tfrac{b}{2a})^2 = \tfrac{b^2 - 4ac}{4a^2} \Rightarrow x + \tfrac{b}{2a} = \pm\sqrt{\tfrac{b^2 - 4ac}{4a^2}}$$

$$\Rightarrow x = -\tfrac{b}{2a} \pm \tfrac{\sqrt{b^2 - 4ac}}{2a} = \tfrac{-b \pm \sqrt{b^2 - 4ac}}{2a} \Rightarrow x = \tfrac{-b \pm \sqrt{b^2 - 4ac}}{2a}.$$

And we call the set of processes above, algebra.

So if a problem is well defined, that is, if it makes sense, we should be able to get it solved the way below:

A problem $\Rightarrow$ ... $\Rightarrow$ ... $\Rightarrow$ the solution, and thus: **the problem $\Rightarrow$ the solution**.

So solving a problem, we put it many different ways so that we can get to the solution.

And that's the way, math runs.

May your math run very well.

Seong R. Kim

B.S. Math. Michigan Tech. Univ.   M.S. Math. Rensselaer Polytechnic Institute

Notes:

This book is about math ideas called logarithms. Why logarithms though?

They are no other than exponents, and are in fact, for your algebra. Doing algebra, you need to know what logarithms are about as well as powers and exponents, how they work so that you can work with them. Why algebra though?

It's simply because you need to *solve problems*. Algebra connects problems to solutions. Algebra only can in fact, actually get you the ones you are always busy finding when taking exams, tests, and quizzes, and doing homework, too, of course. You've got to do algebra to get the very one you want so that you can put it down on *your answer sheet*. With algebra skill, together with your creativity, you can actually solve problems.

And this book is in fact, for your skill of algebra, and you will grow it through examples. Some examples may look too easy or too hard. It all depends on your skill of algebra. Whatever your skill may be though, you can grow yours if you follow the steps in each example. Each is detailed so that you can learn those tools fast, and increase your caliber quickly as well as properly.

And this book explains what logarithms are about and how to manipulate them, that is, how to change or alter, convert, or modify those so that you can come up with the ones that you need. The ones are solutions, of course. And that's what this book is about.

This book does not just explain though. But it helps follow steps to the solutions, too, and thus, helps you do calculations with logarithms as well as powers and exponents so that you can actually do the calculation work doing those manipulations above.

With strong algebra, you can learn things in math fast, and can do problems very well, too, of course. And this book is for your skill of log algebra, which is on expressions with logarithms or those with powers. And there are two more books related to this book, and the two are as follows:

ALGEBRA EXAMPLES POWERS AND LOGARITHMS 1

ALGEBRA EXAMPLES POWERS AND LOGARITHMS 3

And also, all the basics on powers and logarithms and all the ideas contained in this book and the two books above are covered in one book, too. And the book is as follows:

ALGEBRA EXAMPLES POWERS AND LOGARITHMS

So either way, the books will get you not only powers and logarithms but enhancement of your algebra, too. You will thus, soon be able to control powers and logarithms, that is, change or alter, convert, or modify those math expressions so that you can get to the solutions fast. And you will learn them all through examples detailed so that your math can run not only properly but fast enough, too.

Contents

In POWERS AND LOGRAITHMS 2

The Preview of the Contents

In POWERS AND LOGARITHMS 3

The Preview of the Contents

In POWERS AND LOGARITHMS 1

$$(x + y)^2 = x^2 + 2xy + y^2.$$

$$(x + y)^3 = x^3 + 3x^2y + 3xy^2 + y^3.$$

$$(x + y)(x - y) = x^2 - y^2.$$

$$(x + y)(x^2 - xy + y^2) = x^3 + y^3.$$

$$(x^2 + xy + y^2)(x^2 - xy + y^2) = x^4 + x^2y^2 + y^4.$$

$$(x + a)(x + b) = x^2 + (a + b)x + ab.$$

$$(ax + b)(cx + d) = acx^2 + (ad + bc)x + bd.$$

$$(x + a)(x + b)(x + c) = x^3 + (a + b + c)x^2 + (ac + bc + ca)x + abc.$$

$$(a + b + c)^2 = a^2 + b^2 + c^2 + 2(ab + bc + ca).$$

$$(a + b + c)(a^2 + b^2 + c^2 - ab - bc - ca) = a^3 + b^3 + c^3 - 3abc.$$

Suppose both a and $b \neq 0$, and both m and n are integers. Then, we get:

0. $a^m a^n = a^{m+n}$

1. $a^m / a^n = \dfrac{a^m}{a^n} = a^{m-n}$

2. $(a^m)^n = a^{mn}$

3. $(ab)^n = a^n b^n$

4. $(a/b)^n = \left(\dfrac{a}{b}\right)^n = a^n / b^n = \dfrac{a^n}{b^n}$

Suppose both a and $b > 0$, and m and n both are integers nonzero. Then, we get:

0.1. $a^{\frac{1}{n}} b^{\frac{1}{n}} = (ab)^{\frac{1}{n}}$.

1.1. $\dfrac{a^{\frac{1}{n}}}{b^{\frac{1}{n}}} = \left(\dfrac{a}{b}\right)^{\frac{1}{n}}$.

2.1. $(a^{\frac{1}{n}})^m = (a^m)^{\frac{1}{n}}$.

3.1. $(a^{\frac{1}{n}})^{\frac{1}{m}} = a^{\frac{1}{mn}} = (a^{\frac{1}{m}})^{\frac{1}{n}}$.

3.2. $(a^{mp})^{\frac{1}{np}} = (a^m)^{\frac{1}{n}}$, where p is a nonzero integer.

1.   Suppose M, N, and $b > 0$, but $b \neq 1$, and we have: $A = \log_b M$, and $B = \log_b N$. Then, we get: $A - B = \log_b M - \log_b N = \log_b \frac{M}{N}$.

2.   Suppose that M and $b > 0$, but $b \neq 1$, and that we have: $E = \log_b M$. Then, we get: $PE = P \log_b M = \log_b M^P$.

3.   Suppose that a, b, C, and $D > 0$, but a and $b \neq 1$, and that we have: $\log_a C = \log_b D$. Then, we get: $\log_a C = \log_b D = \log_{ab} CD$.

4.   Suppose that a, b, C, and $D > 0$, but a and $b \neq 1$, and that we have: $\log_a C = \log_b D$. Then, we get: $\log_a C = \log_b D = \log_{\frac{a}{b}} \frac{C}{D} = \log_{\frac{b}{a}} \frac{D}{C}$.

5. $\log_b b = 1$, and $\log_b 1 = 0$.

6. $\log_b A = \dfrac{\log_c A}{\log_c b}$.

7. $\log_b A = \dfrac{1}{\log_A b}$.

Note:

The drawings or graphs in this book are not exact, and are approximate or conceptual ones.

$\in$	"$a \in B$" means that a belongs to B. "p, q, and $r \in W$" means that p, q, and r belong to W.
$\Rightarrow$	"$A \Rightarrow B$." means that A implies B.
$\equiv$	$A \equiv B$ means that A and B are identical to each other.
$\neq$	$A \neq B$ means that A is not equal to B.
$|A|$	The magnitude of A. For instance, $|-1| = |1| = 1$.
$\therefore$	Therefore
$\Leftrightarrow$	"$A \Leftrightarrow B$" means "If A then B." and "If B then A." We can read $A \Leftrightarrow B$ as "A if and only if B." In such a case, we can say that $A = B$.
Δx and Δy	Suppose that (x_1, y_1) and (x_2, y_2) are two points in the x-y plane. Then, we get either of the two below. $\Delta x = x_2 - x_1$, and $\Delta y = y_2 - y_1$. $\Delta x = x_1 - x_2$, and $\Delta y = y_1 - y_2$.

Distance Formula

Suppose that d is the distance between two points (x_1, y_1) and (x_2, y_2) in the x-y plane. Then, we get $d^2 = (\Delta x)^2 + (\Delta y)^2$.

6. What is a logarithm?

A logarithm is no other than an exponent, and is often just called a log.
What exponent then, is it?

Exponents have to do with powers. So let's begin with a power.

Suppose for instance, the base of a power is b, and the exponent is x.
Then, we can indicate the power by b^x.

So expressing a power, we can put it in terms of the base and the exponent.

What if however, we want to express the exponent, to which a specific base is raised to equal a particular number, and we want to show the base and the particular number?

For instance, assuming $A = b^x$, how can we put the exponent x in terms of the base b and A, which is the value of the power b^x, and thus, is the particular number we get raising the base b to the exponent x?

So in short, if $A = b^x$, how can we express x in terms of b and A?

Let's begin with some examples.
So for instance, what is the exponent to which a base 10 is raised to equal 1000?

Since 1000 can be said to be 10 to the third power, that is, we can set: $1000 = 10^3$, it is 3.
For another instance, what is the exponent to which a base 2 is raised to equal 16?

Since 16 can be the fourth power of 2, that is, $16 = 2^4$, so it is 4.
Besides, 16 can be put in 4^2, too, so 16 can be called the second power of 4, also.
So for another instance, what is the exponent to which 9 is raised to equal 27?

We have: $27 = 3^3$, and $9 = 3^2$.
So we can set: $3^3 = (3^2)^{\frac{3}{2}} = 9^{\frac{3}{2}}$, and thus, we can get: $27 = 9^{\frac{3}{2}}$.
Therefore, the exponent is $\frac{3}{2}$, so 27 can be called 9 to the three halves power, too.
Besides, we can get 27 raising another base to some other power, too.
Such as what base and power?

We can have: $27 = 3^3$, and $81 = 3^4$, too.
So we can set: $3^3 = (3^4)^{\frac{3}{4}} = 81^{\frac{3}{4}}$, too, and thus, can get: $27 = 81^{\frac{3}{4}}$, also. So the exponent
can be $\frac{3}{4}$, then the base is 81, so 27 can be said to be 81 to the three quarters power, too.

Besides, 27 can still be another base to some other power, too.
That is to say that we can put 27 in terms of another base and another exponent.

In other words, expressing the value of 27, we can use another power where the base is
different, and the exponent is different.

Suppose now, x is an exponent to which a base b is raised to equal a number A.
Then, we can express the exponent x in terms of the number A and the base b. How?

We can do so by means of a special notation, and we call it 'logarithmic notation', often
simply called 'log notation'. And using the *log notation*, we can say we use the *log form*.

So using the log notation, we use the log form, and can specify an exponent in terms of a particular base and the number we get raising the base to the exponent.

So for instance, if the particular base is 2, and the number we get is 8, we can show the exponent 3 putting 2 and 8 in the log form.   How then, do we call the number 8?

The number 8 is called an *antilogarithm*, usually just called an *antilog*, for short.
So expressing an exponent in terms of a base and a number, we call the number an antilogarithm, often simply called an antilog.

Using thus, the log notation, we can put the exponent 3 in terms of the base 2 and the number 8, and we call 8 the antilogarithm.

And we know: $8 = 2^3$. So we can use 2^3 as the antilogarithm, too.

That is to say that using the log form, we can express the exponent 3 in terms of not only 2 and 8 but also 2 and 2^3, which is a power.

So in the log notation, the antilog, that is, the antilogarithm can be either an ordinary number as 49 or a power as 7^2.   What then, is a log, that is, a logarithm?

A log is another name for an exponent, and thus, is an exponent.
As indicated above however, it's not just an exponent, and can be expressed in terms of two numbers. One is called the base, and the other is called the antilog.

How then, can we put an exponent in terms of the two numbers, the base and the antilog?

Normally, specifying a logarithm, we use the log notation. And using the log notation, we use a special sign called a <u>log sign</u>, and as the sign, we use '**log**'. Sometimes though, instead of such a sign, we use '**ln**', too, which is called a <u>natural log sign</u>.

So together with a log sign, we can specify an exponent in terms of a base and an antilog, which is the number we get raising the base to the exponent.

And describing a log specifically, we say <u>a log of a number to a base</u>.

And if we say a log of a number to a base, the number is an antilog.

So for instance, saying <u>the log of 49 to 7</u>, we mean 7 is the base, and 49 is the antilog.

More specifically thus, we can say <u>the log of 49 to base 7</u>.

And of course, the **log of 49 to 7** is an **exponent**, which is 2, since $49 = 7^2$.

So we can say that **2** is the **log of 49 to 7**. And in that case, 2 is an exponent.

• And thus, saying *a log of a number to a base*, we mean an *exponent*, to which the base is raised to equal the number. And the number is called the antilog.

• So a **log of a number to a base** is an **exponent** to which the base is raised to equal the number. And the number is called the antilog.

And thus, using log notation, we can express an exponent in a structured manner, and show the exponent using a base and the number we get raising the base to the exponent.

How then, do we use the **log** sign?   In other words, what does the log form look like?

Suppose we get A raising a base b to an exponent x. That is, $A = b^x$.

Then, putting the exponent x in the log form, we set: $x = \log_b A$, which can be read as: <u>x is equal to the logarithm of A to base b</u>.

Normally though, we read it as: x is equal to the **log** of A to b.

And we can read it this way, too: x is equal to the **log** sub b of A.

And also, x is called the value of $\log_b A$, too, which is a log.

So $\log_b A$ is called a log, and the value of it is x, which can be thus, called a log-value.

And thus, a log-value is an exponent, and is the value of the log of a number to a base.

For instance:

$8 = 2^3 \Leftrightarrow 3 = \log_2 8$, which can be read as the log of 8 to base 2. And the log value is 3.

$\sqrt{2} = 2^{\frac{1}{2}} = 2^{0.5} \Leftrightarrow \frac{1}{2} = \log_2 \sqrt{2}$, can be read as the log of $\sqrt{2}$ to base 2. And the value is $\frac{1}{2}$.

$3 = 27^{\frac{1}{3}} \Leftrightarrow \frac{1}{3} = \log_{27} 3$, can be read as the log of 3 to 27. And the value is $\frac{1}{3}$.

$\frac{1}{3} = 81^{-\frac{1}{4}} \Leftrightarrow -\frac{1}{4} = \log_{81} \frac{1}{3}$, which can be read as the log sub 81 of $\frac{1}{3}$. And the value is $-\frac{1}{4}$.

$9.7385 \approx 5^{\sqrt{2}} \Leftrightarrow \sqrt{2} \approx \log_5 9.7385$, which is saying that:

The value of the log sub 5 of 9.7385 is approximately $\sqrt{2}$.

$1385.45573 \approx 10^{\pi} \Leftrightarrow \pi \approx \log_{10} 1385.45573$, which is saying that:

The value of the log sub 10 of 1385.45573 is approximately π, which is the circular ratio, which is approximately 3.141592.

So using log notation, that is, the log form, we can express a particular exponent in a structured manner, where we can show a particular base and the number we get raising the particular base to the particular exponent. Why particular, though?

Many powers can have the same value.

For instance, we can have: $64 = 2^6 = 4^3 = 8^2 \approx 10^{1.806}$, and also, $64 = 4096^{0.5} = 0.5^{-6} = \ldots$

So we can get: $6 = \log_2 64$, $3 = \log_4 64$, $2 = \log_8 64$, $1.806 \approx \log_{10} 64$, $0.5 = \log_{4096} 64$, $\ldots$

And thus, we can put the idea above, the way below, too:

If the value of a power or an antilog remains the same, but the base changes, then the exponent changes, too.

So for instance, assuming $A = b^x$, that is, $x = \log_b A$, and keeping A constant, but changing the value of b, we have to change the value of x, too.

• In short, if $A = b^x$ or $x = \log_b A$, and A remains the same, but b changes, x changes, too.

Thus, given a particular base and an antilog, we can track down a particular exponent corresponding to the particular base.

On the other hand, keeping the base constant in a log, if we vary the antilog, what happens to the exponent?

That is, if $x = \log_b A$, and we make A change keeping b constant, what happens to x?

The exponent x changes, too, of course.   So what?

We can set up a correlation between an antilog and an exponent.
That is to say that we can make a function, where the value of an exponent changes in accordance with the change of the value of an antilog.

Therefore, by means of such a function, we can keep track of an exponent (a variable exponent), which varies as the antilog changes with respect to a particular base fixed.

Such a function is called a **log function**, which is a function of antilog. So each input of a log function is an antilog, and each output of a log function is an exponent.

On the other hand, we can set up another function where the reverse can be achieved.

Keeping the base constant in a power, if we vary the exponent, what happens to the value of the power?
That is, if $A = b^x$, and we make x change keeping b constant, what happens to A?

The value of A changes, too, of course.

So we can set up a correlation between an exponent and a power.
That is to say that we can make a function, where the value of a power changes in accordance with the change of the value of an exponent.

Therefore, by means of such a function, we can keep track of (the value of) a power, which varies as the exponent changes with respect to a particular base fixed.

Such a function is called an **exponential function**, which is a function of exponent. So each input of an exponential function is an exponent, and each output of an exponential function is the value of a power.

Thus, a **log function** is taken as the **inverse** of an **exponential function**, and vice versa.

(You can find more details on functions themselves, in the book **ALGEBRA EXAMPLES BASIC FUNCTIONS.**)

Now, what is the difference between the value of a power and the value of an antilog?

If we get a number raising a base to an exponent, what can we call the number?

We can call the number two different ways:

- Putting the number in term of the base and the exponent, we call the number a power.

- Using log notation, we can put an exponent in terms of a base and the number we get raising the base to the exponent. Then, we can call the number an antilog.

So if we get a number raising a base to an exponent, we can call the number a power or an antilog. So an antilog can be not only a number as 8 but a power as 2^3, too.

Thus, for instance, we can have: $3 = \log_2 8$, and can put it this way, too: $3 = \log_2 2^3$.

And putting the idea above more specifically, we can say that:

- Getting a number raising a base to an exponent, and putting the number in power notation, that is, the power form, we call the number a power.

- Getting a number raising a base to an exponent, and putting the exponent in log notation, that is, the log form, we call the number an antilog.

That is to say that if $A = b^x$, we get: $x = \log_b A$, and also, if $x = \log_b A$, we get: $A = b^x$, too.

In other words, we have: $A = b^x \Rightarrow x = \log_b A$, and also, $x = \log_b A \Rightarrow A = b^x$, too.

- In short: $A = b^x \Leftrightarrow x = \log_b A$, which is read as: $A = b^x$ if and only if $x = \log_b A$.

So for instance, we can have:

$9 = 3^2 \Rightarrow 2 = \log_3 9$, and also, $2 = \log_3 9 \Rightarrow 9 = 3^2$.
In short, $9 = 3^2 \Leftrightarrow 2 = \log_3 9$.

$0.001 = 0.1^3 \Rightarrow 3 = \log_{0.1} 0.001$, and also, $3 = \log_{0.1} 0.001 \Rightarrow 0.001 = 0.1^3$.
In short, $0.001 = 0.1^3 \Leftrightarrow 3 = \log_{0.1} 0.001$.

What then, is a log value?

It's the value of a log, and is in fact, an exponent.

So for instance, 3 is the value of $\log_2 8$.
And since $\sqrt{2} = 2^{\frac{1}{2}}$, $\frac{1}{2}$ is an exponent, and is the value of $\log_2 \sqrt{2}$.

And we have: $3 = \log_2 8 = \log_{0.1} 0.001 = \ldots$
So many logs looking different can have the same value, and thus, can be the same.

Samples on Powers & Logarithms

$8 = 2^3 \Leftrightarrow 3 = \log_2 8$ $243 = 3^5 \Leftrightarrow 5 = \log_3 243$

$0.125 = 0.5^3 \Leftrightarrow 3 = \log_{0.5} 0.125$ $0.008 = 5^{-3} \Leftrightarrow -3 = \log_5 0.008$

$0.00243 = 0.3^5 \Leftrightarrow 5 = \log_{0.3} 0.00243$ $3125 = 0.2^{-5} \Leftrightarrow -5 = \log_{0.2} 3125$

$16 = 256^{0.5} \Leftrightarrow 0.5 = \log_{256} 16$ $0.0625 = 256^{-0.5} \Leftrightarrow -0.5 = \log_{256} 0.0625$

$4\sqrt{2} = \sqrt{32} = (2^5)^{\frac{1}{2}} = 2^{\frac{5}{2}} \Leftrightarrow \frac{5}{2} = \log_2 4\sqrt{2} = \log_2 \sqrt{32} = \log_2 (2^5)^{\frac{1}{2}} = \log_2 2^{\frac{5}{2}}$

$2^2 2^{\frac{1}{2}} = 2^{2+\frac{1}{2}} = 2^{\frac{5}{2}} \Leftrightarrow \frac{5}{2} = \log_2 2^2 2^{\frac{1}{2}} = \log_2 2^{2+\frac{1}{2}} = \log_2 2^{\frac{5}{2}}$

$2^2 \div 2^{\frac{1}{2}} = 2^{2-\frac{1}{2}} = 2^{\frac{3}{2}} \Leftrightarrow \frac{3}{2} = \log_2 2^2 \div 2^{\frac{1}{2}} = \log_2 2^{2-\frac{1}{2}} = \log_2 2^{\frac{3}{2}}$

Note that we have: $\log_2 2^2 \div 2^{\frac{1}{2}} = \log_2 (2^2 \div 2^{\frac{1}{2}})$.

$$(2^2)^{\frac{1}{2}} = 2^{2 \cdot \frac{1}{2}} = 2^1 = 2 \Leftrightarrow 1 = \log_2 (2^2)^{\frac{1}{2}} = \log_2 2^{2 \cdot \frac{1}{2}} = \log_2 2^1 = \log_2 2$$

$$\sqrt{2^2} = (2^2)^{\frac{1}{2}} = 2 \Leftrightarrow 1 = \log_2 \sqrt{2^2} = \log_2 (2^2)^{\frac{1}{2}} = \log_2 2$$

$$6^{0.8} = 6^{\frac{8}{10}} = 6^{\frac{4}{5}} = \sqrt[5]{6^4} \Leftrightarrow 0.8 = \log_6 6^{0.8} = \log_6 6^{\frac{8}{10}} = \log_6 6^{\frac{4}{5}} = \log_6 \sqrt[5]{6^4}$$

$$6^{0.8} = (2 \cdot 3)^{0.8} = 2^{0.8} 3^{0.8} \Leftrightarrow 0.8 = \log_6 6^{0.8} = \log_6 (2 \cdot 3)^{0.8} = \log_6 2^{0.8} 3^{0.8}$$

$$\left(\frac{2}{3}\right)^{\frac{1}{2}} = \sqrt{\frac{2}{3}} = \frac{\sqrt{2}}{\sqrt{3}} = \frac{2^{\frac{1}{2}}}{3^{\frac{1}{2}}} \Leftrightarrow \frac{1}{2} = \log_{\frac{2}{3}} \left(\frac{2}{3}\right)^{\frac{1}{2}} = \log_{\frac{2}{3}} \sqrt{\frac{2}{3}} = \log_{\frac{2}{3}} \frac{\sqrt{2}}{\sqrt{3}} = \log_{\frac{2}{3}} \frac{2^{\frac{1}{2}}}{3^{\frac{1}{2}}}$$

Note that we have: $\log_{\frac{2}{3}} \dfrac{2^{\frac{1}{2}}}{3^{\frac{1}{2}}} = \log_{\frac{2}{3}} \left(\dfrac{2^{\frac{1}{2}}}{3^{\frac{1}{2}}}\right)$.

$$\left(\frac{2}{3}\right)^{\frac{2}{3}} = \sqrt[3]{\left(\frac{2}{3}\right)^2} = \sqrt[3]{\frac{2^2}{3^2}} = \frac{\sqrt[3]{2^2}}{\sqrt[3]{3^2}} = \frac{2^{\frac{2}{3}}}{3^{\frac{2}{3}}}$$

$$\Leftrightarrow \frac{2}{3} = \log_{\frac{2}{3}} \left(\frac{2}{3}\right)^{\frac{2}{3}} = \log_{\frac{2}{3}} \sqrt[3]{\left(\frac{2}{3}\right)^2} = \log_{\frac{2}{3}} \sqrt[3]{\frac{2^2}{3^2}} = \log_{\frac{2}{3}} \frac{\sqrt[3]{2^2}}{\sqrt[3]{3^2}} = \log_{\frac{2}{3}} \frac{2^{\frac{2}{3}}}{3^{\frac{2}{3}}}$$

$$\left(\frac{2}{3}\right)^{-\frac{5}{3}} = \sqrt[3]{\left(\frac{2}{3}\right)^{-5}} = \sqrt[3]{\left(\frac{3}{2}\right)^{5}} = \sqrt[3]{\frac{3^5}{2^5}} = \frac{\sqrt[3]{3^5}}{\sqrt[3]{2^5}} = \frac{3^{\frac{5}{3}}}{2^{\frac{5}{3}}}$$

$$\Leftrightarrow -\frac{5}{3} = \log_{\frac{2}{3}}\left(\frac{2}{3}\right)^{-\frac{5}{3}} = \log_{\frac{2}{3}}\sqrt[3]{\left(\frac{2}{3}\right)^{-5}} = \log_{\frac{2}{3}}\sqrt[3]{\left(\frac{3}{2}\right)^{5}} = \log_{\frac{2}{3}}\sqrt[3]{\frac{3^5}{2^5}} = \log_{\frac{2}{3}}\frac{\sqrt[3]{3^5}}{\sqrt[3]{2^5}} = \log_{\frac{2}{3}}\frac{3^{\frac{5}{3}}}{2^{\frac{5}{3}}}$$

$$\left(\frac{2}{3}\right)^{-0.1} = \frac{2^{-0.1}}{3^{-0.1}} = \frac{\sqrt[10]{2^{-1}}}{\sqrt[10]{3^{-1}}} = \frac{\sqrt[10]{\frac{1}{2}}}{\sqrt[10]{\frac{1}{3}}} = \sqrt[10]{\frac{\frac{1}{2}}{\frac{1}{3}}} = \sqrt[10]{\frac{3}{2}}$$

$$\Leftrightarrow -0.1 = \log_{\frac{2}{3}}\left(\frac{2}{3}\right)^{-0.1} = \log_{\frac{2}{3}}\frac{2^{-0.1}}{3^{-0.1}} = \log_{\frac{2}{3}}\frac{\sqrt[10]{2^{-1}}}{\sqrt[10]{3^{-1}}} = \log_{\frac{2}{3}}\frac{\sqrt[10]{\frac{1}{2}}}{\sqrt[10]{\frac{1}{3}}} = \log_{\frac{2}{3}}\sqrt[10]{\frac{\frac{1}{2}}{\frac{1}{3}}} = \log_{\frac{2}{3}}\sqrt[10]{\frac{3}{2}}$$

$$\left(\frac{2}{3}\right)^{-0.1} = \left(\left(\frac{2}{3}\right)^{-1}\right)^{0.1} = \left(\frac{3}{2}\right)^{0.1} = \frac{3^{0.1}}{2^{0.1}} = \frac{\sqrt[10]{3}}{\sqrt[10]{2}} = \sqrt[10]{\frac{3}{2}}$$

$$\Leftrightarrow -0.1 = \log_{\frac{2}{3}}\left(\frac{2}{3}\right)^{-0.1} = \log_{\frac{2}{3}}\left(\left(\frac{2}{3}\right)^{-1}\right)^{0.1} = \log_{\frac{2}{3}}\left(\frac{3}{2}\right)^{0.1} = \log_{\frac{2}{3}}\frac{3^{0.1}}{2^{0.1}} = \log_{\frac{2}{3}}\frac{\sqrt[10]{3}}{\sqrt[10]{2}} = \log_{\frac{2}{3}}\sqrt[10]{\frac{3}{2}}$$

7. Definitions for Logarithms

To begin with, a logarithm is called a log, for short, and is another name for an exponent.
So in short, a log is an exponent.   What then, is a log-value?

Taking a log of a number, and then, taking the value of the log, we get a log-value.
For instance, $\log_2 8$ is a log, and the value of it is 3, which is a log-value.
So we set: $3 = \log_2 8$, where 2 is called the base, and 8 is called the antilog.

So more specifically, a log-value is the value of the log of a number to a particular base.
Taking thus, the value of the log of a number to a base, we get the log-value to the base.

So a log-value depends on the base to which we take a log of the number. For instance, the log of 81 to base 3 is 4, but the log of 81 to base 9 is 2, so in this case, 4 is the log-value to base 3, and 2 is the log-value to base 9.   So what is a log-value?

It's an exponent that we use expressing a number in a power of a particular base.
So a log value is the value of a log, and is a number meaning an exponent.

And a log value is real, but <u>the antilog in a log is positive only</u>.
In other words, a log-value belongs to a set of all real numbers, and the antilog in a log belongs to a set of all positive numbers.
That is, all real numbers need to be allowed to be a log-value, and all positive numbers need to be allowed to be the antilog in a log.
And <u>the base in a log</u> has to be <u>positive but unequal to 1</u>.

What then, is the definition for logarithms?

Suppose now, R is a set of all real numbers, and A, b, and $x \in R$.
Suppose also, A and b both > 0, but $b \neq 1$.

Then, $A = b^x \Leftrightarrow x = \log_b A$.

The entire set of all the three statements above can be called the definition for logarithms.

We can have two kinds in definitions. One is a full definition, and the other is a short definition. The entire set of statements above can be called the full definition for logs, and we can take as the short definition the last statement where "$A = b^x \Leftrightarrow x = \log_b A$".

And we often use the short version, because it is convenient for a quick reference.

Keep in mind though, the full definition, too.
That's because the two suppositions in the full version are intrinsic to logarithms, and in particular, the first of the two is often a vital condition we need to satisfy solving many problems on logarithms.

So when solving problems with logs, we want to check to see if the solutions meet the condition. And in fact, we can readily solve many of log problems by means of the condition, too. What then, is the condition?

The full definition has the condition, and is shown below:

Suppose that:
R is a set of all real numbers, and A, b, and $x \in R$.
$A > 0$, and $b > 0$, but $b \neq 1$.
Then, $A = b^x \Leftrightarrow x = \log_b A$.

In the full definition above, the statement that $\underline{A > 0, \text{ and } b > 0 \text{ but } b \neq 1}$ is the condition.

And in the statement above:

- "$A > 0$." is saying that *the antilog in a log* is *positive*.

In other words, the antilog in a log belongs to a set of all positive numbers.

So numbers <u>positive only</u> can be assigned to the <u>antilog A</u> in $\log_b A$, and thus, the antilog cannot be negative or 0.

Next, in the statement above:

- "$b > 0$ but $b \neq 1$" is saying that <u>the base in a log</u> is positive, but cannot be 1.

So if we can get: $\log_b A$, that is, if we can take the log of A to b (or the log sub b of A), it has to be the case where: A and $b > 0$, yet $b \neq 1$. And the reverse is true, too.

So if A and $b > 0$, yet $b \neq 1$, we can get: $\log_b A$, that is, we can take the log sub b of A.

Thus, the statement "A and $b > 0$ but $b \neq 1$." is a favorite with many examiners including probably your teacher, too. So keep in mind that *in a log*, the antilog and the base have to be positive, but the base cannot be 1.

Why in the definition though, do we have: $b > 0$ but $b \neq 1$?

Suppose first, $A = b^x$, and $\underline{b < 0}$.

Then, we **cannot** use all real numbers as the exponent x.   Why not?

If $b = -2$ and $x = 0.5$, we get: $A = b^x = (-2)^{0.5} = \sqrt{-2}$, which is however, not real.

So if $b < 0$, A can be in trouble, and thus, it can be the case where we cannot get $\log_b A$.

So we want the base b to be positive only. Then of course, A can be positive only, too.

18

If however, $b > 0$, we can always get: $A = b^x$ no matter what real number x may be, so in turn, we can always get: $x = \log_b A$. Thus, if $b > 0$, we can use all real numbers as x in b^x.

What then, about the case where the base is 0 or 1?

If the base b is 0, we cannot use any number negative as the exponent x, because it causes 0 to be a denominator. For instance, $0^{-2} = 1/0^2 = 1/0$, which is not allowed.

And in fact, if $x < 0$, $0^x = 1/0^{-x} = 1/0$, which is not allowed.

So if $b = 0$, and $x < 0$, we cannot get: $A = b^x$, and thus, in turn, we cannot get: $x = \log_b A$.

And what's even worse is that we cannot take the log sub 0 of any number other than 1. Why not?

Suppose for instance, $A = 2$, and $b = 0$. Then by definition, we get: $x = \log_b A = \log_0 2$, which can be however, no number, because there is no x for which $0^x = 2$.

And thus, 0 cannot be the base.

And next, if the base b is 1, we just get: $A = b^x = 1^x = 1$, no matter what real number the exponent x may be.

So taking the log of 1 to 1, that is, the log sub 1 of 1, we do not get a particular number as the value of the log. That is, $\log_1 1$ can be any number.   How come?

We can get: $1 = 1^x$ for all x, and thus, for all x, we can get: $x = \log_1 1$.

And what's even worse is that we cannot take the log sub 1 of any number other than 1. Why not?

Suppose for instance, $A = 2$, and $b = 1$. Then by definition, we get: $x = \log_b A = \log_1 2$, which can be however, no number, because there is no x for which $1^x = 2$.

So gathering up the threads, we can say that the *base* has to be *positive* but *unequal to* 1.

Specifically, assuming b is the base in a log, and keeping the definition happy, we need to have: $0 < b < 1$, or $1 < b$.

We have already covered the details on the problems with bases negative in the sections for **Problem Bases**.

In the next section though, we will cover more details on the bases used in logarithms.

That's because even if the base is positive, and is not 1, the base can still matters if it is used in a logarithm. And it does very much so.

Now, getting back to the full definition for logs, we have:

Assuming R is a set of all real numbers, A, b, and $x \in R$, $A > 0$, and $b > 0$, but $b \neq 1$, we get: $A = b^x \Leftrightarrow x = \log_b A$.

So keep in mind that <u>the base in a log</u> has to be <u>positive but unequal to 1</u>.

And thus, a log-value is real, and the antilog is positive.

In other words, a log-value belongs to a set of all real numbers, and the antilog belongs to a set of all positive numbers.

And usually, we work with the short version below:

$A = b^x \Leftrightarrow x = \log_b A$ (where A and b both > 0, but $b \neq 1$).

What do we mean by though, defining a log?

Defining a log, we make a log.
And of course, taking a log of a number to a base, we make a log, too.
And making a log, we define a log, too.

So for instance, defining a log where the base is 2 and the antilog is 8, we make a log, which is: $\log_2 8$, which can be therefore, called a log definition.

So a log definition is the definition of a log, and is a log, but is not the definition for logs.

The definition for logs explains or shows what logs are, and can be: $A = b^x \Leftrightarrow x = \log_b A$.

And a log definition shows or specifies a particular log, and for instance, can be $\log_3 9$.

Examples 1 on Definitions for Logarithms

Find x in each case below:

0. $\log_x C = \frac{3}{5}$ where $C = 2\sqrt{2}$.

1. $\log D = -2$ where $D = 0.02 \log_{16} x$.

2. $\log_a (\log_b (\log_c x)) = 0$ where $a = 3.97$, $b = 3$, and $c = 5$.

3. $\log_a (\log_b (\log_c x)) = 2$ where $a = 0.1$, $b = 3^{100}$, and $c = 5$.

Suggestions or Solutions
To the Examples 1 on Definitions for Logarithms

0. $\log_x C = \frac{3}{5}$ where $C = 2\sqrt{2}$.

First, by the definition for logs, we get: $x^{\frac{3}{5}} = C$.

And we can have: $C = 2\sqrt{2} = 2^{\frac{3}{2}}$, too. Thus, we get: $x^{\frac{3}{5}} = C = 2^{\frac{3}{2}}$.

Next, since the base is unknown, we may want to make the exponents the same.

$$\frac{3}{2} = \frac{3}{5} \cdot \frac{5}{2} = \frac{5}{2} \cdot \frac{3}{5} \Rightarrow 2^{\frac{3}{2}} = 2^{\frac{5}{2} \cdot \frac{3}{5}} = (2^{\frac{5}{2}})^{\frac{3}{5}}.$$

Thus, we get: $x^{\frac{3}{5}} = 2^{\frac{3}{2}} \Rightarrow x^{\frac{3}{5}} = (2^{\frac{5}{2}})^{\frac{3}{5}} \Rightarrow x = 2^{\frac{5}{2}}$, which is $\sqrt{2^5} = \sqrt{32} = 4\sqrt{2}$.

In short:

$$x^{\frac{3}{5}} = C = 2\sqrt{2} = 2^{\frac{3}{2}} = 2^{\frac{5}{2} \cdot \frac{3}{5}} = (2^{\frac{5}{2}})^{\frac{3}{5}} = (\sqrt{2^5})^{\frac{3}{5}} \Rightarrow x = \sqrt{32} = 4\sqrt{2}.$$

1. $\log D = -2$ where $D = 0.02 \log_{16} x$.

By the definition for logs, we get: $\log D = -2 \Leftrightarrow 10^{-2} = D$.
So we get: $D = 0.01$.
Thus, we get: $D = 0.02 \log_{16} x \Rightarrow 0.01 = 0.02 \log_{16} x \Rightarrow 1/2 = \log_{16} x$, and then again, by the definition, we get: $x = 16^{0.5} = 4$.

In short:

By the definition for logs, we get: $\log D = -2 \Leftrightarrow 10^{-2} = D = 0.01$.

So we get: $D = 0.02 \log_{16} x \Rightarrow 0.01 = 0.02 \log_{16} x \Rightarrow 1/2 = 0.5 = \log_{16} x \Leftrightarrow x = 16^{0.5} = 4$.

2. $\log_a (\log_b (\log_c x)) = 0$ **where** $a = 3.97$, $b = 3$, **and** $c = 5$.

By the definition for logs, we get: $\log_b (\log_c x) = a^0 = 1$.

Then again, by the definition, we get: $\log_c x = b^1 = b$, which is 3.

Then, by the definition again, we get: $x = c^3$ where $c = 5$.

Thus, we get: $x = 5^3$.

In short:

$\log_a (\log_b (\log_c x)) = 0 \Leftrightarrow \log_b (\log_c x) = a^0 = 1 \Leftrightarrow \log_c x = b^1 = 3 \Leftrightarrow x = c^3 = 5^3 = 125$.

3. $\log_a (\log_b (\log_c x)) = 2$ **where** $a = 0.1$, $b = 3^{100}$, **and** $c = 5$.

By the definition for logs, we get: $\log_b (\log_c x) = a^2 = (0.1)^2 = 0.01$.

Then again, by the definition, we get: $\log_c x = b^{0.01} = (3^{100})^{0.01} = 3$.

Then, by the definition again, we get: $x = c^3$ where $c = 5$.

Thus, we get: $x = 5^3$.

In short:

$\log_a (\log_b (\log_c x)) = 2 \Leftrightarrow \log_b (\log_c x) = a^2 = 0.01$

$\Leftrightarrow \log_c x = b^{0.01} = (3^{100})^{0.01} = 3 \Leftrightarrow x = c^3 = 5^3 = 125$.

Examples 2 on Definitions for Logarithms

Find the value of x in each case below:

0. $x = \log_3 81$

1. $x = \log_{3\sqrt{2}} 324$

2. $x = \log_3 9\sqrt{3}$

3. $x = \log_{0.2} 125$

4. $x = \log_{0.4} 15.625$

5. $x = \log_{0.3} 0.027\sqrt{0.3}$

6. $x = \log_{16}(\sqrt{7+4\sqrt{3}} + \sqrt{7-4\sqrt{3}})$

7. $x = \log_{12}(\sqrt{5+2\sqrt{6}} + \sqrt{5-2\sqrt{6}})$

8. $x = \log_4(\sqrt{3+\sqrt{5}} - \sqrt{3-\sqrt{5}})$

9. $\log_x 3\sqrt{3} = \frac{2}{3}$

A. $\log_{\sqrt{x}} 3\sqrt{3} = \frac{5}{2}$

B. $\log_{\sqrt{x}} 2\sqrt{3} = -\frac{1}{2}$

C. $\log_x 125 = 5$

D. $\log_{\sqrt{x}} 15.625 = -3$

E. $x = \log_{0.5} 8$

F. $\log_2 3x = 2$

G. $\log_3 3x = 9$

H. $\log_2 (\log_3 x) = 1$

I. $\log_{0.2} (\log_{0.3} x) = 2$

J. $\log_{12} (\log_3 x^2) = 3$

K. $\log_3 (\log_{0.2} x^2) = -2$

L. $\log_{0.2} (\log_{12} x^{-2}) = 3$

M. $\log_2 (\log_x 16) = 3$

N. $\log_2 (\log_{\sqrt{x}} 8) = -3$

O. $\log_2 (\log_{x^2} 3\sqrt{3}) = -2$

P. $\log_2 (\log_{x^{0.3}} 4) = -1$

Q. $\log_2 \{\log_3 (\log_5 x)\} = 2$

Suggestions or Solutions
To the Examples 2 on Definitions for Logarithms

The definition for logs is: $A = b^x \Leftrightarrow x = \log_b A$.

So using the definition, we can get the solutions the way below.

0. $x = \log_3 81 \Leftrightarrow 81 = 3^x$.   And we have: $81 = 3^4$.   So we get: $x = 4$.

1. $x = \log_{3\sqrt{2}} 324 \Leftrightarrow 324 = (3\sqrt{2})^x$.

And we have: $324 = 2 \cdot 162 = 2 \cdot 2 \cdot 81 = 2^2 3^4 = (\sqrt{2})^4 3^4 = (3\sqrt{2})^4$.   So we get: $x = 4$.

2. $x = \log_3 9\sqrt{3} \Leftrightarrow 9\sqrt{3} = 3^x$.   And we have: $9\sqrt{3} = 3^2 3^{0.5} = 3^{2.5}$.   So we get: $x = 2.5$.

3. $x = \log_{0.2} 125 \Leftrightarrow 125 = 0.2^x$.   And we have: $0.2 = 1/5 = 5^{-1}$, and $125 = 5 \cdot 25 = 5^3$.

So we get: $125 = 0.2^x \Rightarrow 5^3 = (5^{-1})^x = 5^{-x} \Rightarrow 5^3 = 5^{-x} \Rightarrow x = -3$.

4. $x = \log_{0.4} 15.625 \Leftrightarrow 15.625 = 0.4^x$.

And we have: $0.4 = 2/5 = 2 \cdot 5^{-1}$, and $15.625 = 15625 \cdot 10^{-3}$.

So we get: $0.4^x = (2 \cdot 5^{-1})^x$.

And we have: $15625 = 5 \cdot 3125 = 5^2 \cdot 625 = 5^2 \cdot 25^2 = 5^6$, and $10^{-3} = 2^{-3} 5^{-3}$.

So we get: $15.625 = 15625 \cdot 10^{-3} = 5^6 2^{-3} 5^{-3} = 2^{-3} 5^3 = (2^3 5^{-3})^{-1} = (2 \cdot 5^{-1})^{-3}$.

Thus, we get: $15.625 = 0.4^x \Rightarrow (2 \cdot 5^{-1})^{-3} = (2 \cdot 5^{-1})^x \Rightarrow x = -3$.

5. $x = \log_{0.3} 0.027\sqrt{0.3} \Leftrightarrow 0.027\sqrt{0.3} = 0.3^x$.

And we have: $0.027\sqrt{0.3} = 0.3^3 0.3^{0.5} = 0.3^{3.5}$.   So we get: $x = 3.5$.

6. $x = \log_{16}(\sqrt{7 + 4\sqrt{3}} + \sqrt{7 - 4\sqrt{3}}) \Leftrightarrow \sqrt{7 + 4\sqrt{3}} + \sqrt{7 - 4\sqrt{3}} = 16^x$.

And we have: $7 + 4\sqrt{3} = 7 + 2\sqrt{4 \cdot 3} = 4 + 3 + 2\sqrt{4 \cdot 3} = (\sqrt{4} + \sqrt{3})^2 = (2 + \sqrt{3})^2$.

So by the same token, we get: $7 - 4\sqrt{3} = 7 - 2\sqrt{4 \cdot 3} = 4 + 3 - 2\sqrt{4 \cdot 3} = (2 - \sqrt{3})^2$.

Thus, we get: $\sqrt{7 + 4\sqrt{3}} + \sqrt{7 - 4\sqrt{3}} = 2 + \sqrt{3} + 2 - \sqrt{3} = 4$.

So we get: $4 = 16^x = 4^{2x} \Rightarrow 2x = 1 \Rightarrow x = 1/2$.

7. $x = \log_{12}(\sqrt{5+2\sqrt{6}} + \sqrt{5-2\sqrt{6}}) \Leftrightarrow \sqrt{5+2\sqrt{6}} + \sqrt{5-2\sqrt{6}} = 12^x$.

And we have: $5 + 2\sqrt{6} = 5 + 2\sqrt{3 \cdot 2} = 3 + 2 + 2\sqrt{3 \cdot 2} = (\sqrt{3} + \sqrt{2})^2$.

By the same token, we get: $5 - 2\sqrt{6} = 5 - 2\sqrt{3 \cdot 2} = 3 + 2 - 2\sqrt{3 \cdot 2} = (\sqrt{3} - \sqrt{2})^2$.

So we get: $\sqrt{5+2\sqrt{6}} + \sqrt{5-2\sqrt{6}} = \sqrt{3} + \sqrt{2} + \sqrt{3} - \sqrt{2} = 2\sqrt{3}$.

Thus, we get: $2\sqrt{3} = 12^x$. And we have: $2\sqrt{3} = \sqrt{12} = 12^{0.5}$.

So we get: $12^{0.5} = 12^x \Rightarrow x = 0.5$.

8. $x = \log_4(\sqrt{3+\sqrt{5}} - \sqrt{3-\sqrt{5}}) \Leftrightarrow \sqrt{3+\sqrt{5}} - \sqrt{3-\sqrt{5}} = 4^x$.

And we have: $3 + \sqrt{5} = \frac{1}{2}(6 + 2\sqrt{5}) = \frac{1}{2}(5 + 1 + 2\sqrt{5 \cdot 1}) = \frac{1}{2}(\sqrt{5} + 1)^2$.

By the same token, we get: $3 - \sqrt{5} = \frac{1}{2}(6 - 2\sqrt{5}) = \frac{1}{2}(5 + 1 - 2\sqrt{5 \cdot 1}) = \frac{1}{2}(\sqrt{5} - 1)^2$.

So we get: $\sqrt{3+\sqrt{5}} - \sqrt{3-\sqrt{5}} = \frac{1}{\sqrt{2}}\{(\sqrt{5}+1)-(\sqrt{5}-1)\} = \frac{2}{\sqrt{2}} = \sqrt{2} = 2^{0.5}$.

Thus, we get: $2^{0.5} = 4^x$, which is 2^{2x}.   So we get: $2x = 0.5 \Rightarrow x = 1/4$.

9. $\log_x 3\sqrt{3} = \frac{2}{3} \Leftrightarrow 3\sqrt{3} = x^{\frac{2}{3}}$.

And we have: $3\sqrt{3} = 3^{\frac{3}{2}} = 3^{\frac{3}{2} \cdot 1} = 3^{\frac{3}{2} \cdot \frac{2}{3} \cdot \frac{3}{2}} = 3^{\frac{9}{4} \cdot \frac{2}{3}} = (3^{\frac{9}{4}})^{\frac{2}{3}}$.

So we get: $3\sqrt{3} = x^{\frac{2}{3}} \Rightarrow (3^{\frac{9}{4}})^{\frac{2}{3}} = x^{\frac{2}{3}} \Rightarrow x = 3^{\frac{9}{4}}$.

A. $\log_{\sqrt{x}} 3\sqrt{3} = \frac{5}{2} \Leftrightarrow 3\sqrt{3} = (\sqrt{x})^{\frac{5}{2}}$.

And we have: $(\sqrt{x})^{\frac{5}{2}} = (x^{\frac{1}{2}})^{\frac{5}{2}} = x^{\frac{5}{4}}$. Also, we get: $3\sqrt{3} = 3^{\frac{3}{2}} = 3^{\frac{3}{2}\cdot 1} = 3^{\frac{3}{2}\cdot\frac{4}{5}\cdot\frac{5}{4}} = (3^{\frac{6}{5}})^{\frac{5}{4}}$.

So we get: $(3^{\frac{6}{5}})^{\frac{5}{4}} = x^{\frac{5}{4}} \Rightarrow x = 3^{\frac{6}{5}}$.

B. $\log_{\sqrt{x}} 2\sqrt{3} = -\frac{1}{2} \Leftrightarrow 2\sqrt{3} = (\sqrt{x})^{-\frac{1}{2}} = x^{-\frac{1}{4}}$.

And we have: $2\sqrt{3} = \sqrt{12} = 12^{\frac{1}{2}} = (12^2)^{\frac{1}{4}} = (12^{-2})^{-\frac{1}{4}}$.

So we get: $(12^{-2})^{-\frac{1}{4}} = x^{-\frac{1}{4}} \Rightarrow x = 12^{-2}$.

C. $\log_x 125 = 5 \Leftrightarrow 125 = x^5$.

And we have: $125 = 5^3 = 5^{3\cdot 1} = 5^{3\cdot\frac{1}{5}\cdot 5} = 5^{\frac{3}{5}\cdot 5} = (5^{\frac{3}{5}})^5$.

So we get: $(5^{\frac{3}{5}})^5 = x^5 \Rightarrow x = 5^{\frac{3}{5}}$.

D. $\log_{\sqrt{x}} 15.625 = -3 \Leftrightarrow 15.625 = (\sqrt{x})^{-3} = x^{-\frac{3}{2}}$. And we have:

$15.625 = 15625\cdot 10^{-3} = 5^6 10^{-3} = 5^6(5\cdot 2)^{-3} = 5^3 2^{-3}$

$= (\frac{5}{2})^3 = (\frac{5}{2})^{3\cdot 1} = (\frac{5}{2})^{3\cdot(-\frac{1}{2})(-2)} = (\frac{5}{2})^{(-\frac{3}{2})(-2)} = (\frac{5}{2})^{(-2)(-\frac{3}{2})} = \{(\frac{5}{2})^{(-2)}\}^{-\frac{3}{2}} = (\frac{4}{25})^{-\frac{3}{2}}$.

So we get: $(\frac{4}{25})^{-\frac{3}{2}} = x^{-\frac{3}{2}} \Rightarrow x = \frac{4}{25}$.

E. $x = \log_{0.5} 8 \Leftrightarrow 8 = x^{0.5}$. And we have: $8 = 2^3 = 2^{6\cdot 0.5} = (2^6)^{0.5}$.

So we get: $8 = x^{0.5} \Rightarrow (2^6)^{0.5} = x^{0.5} \Rightarrow x = 2^6$.

F. $\log_2 3x = 2 \Leftrightarrow 3x = 2^2 = 4$.   So we get: $3x = 4 \Rightarrow x = 4/3$.

G. $\log_3 3x = 9 \Leftrightarrow 3x = 3^9$.   So we get: $x = 3^9/3 = 3^{9-1} = 3^8$.

H. $\log_2 (\log_3 x) = 1 \Leftrightarrow \log_3 x = 2^1 = 2$.

And by the definition again, we can get: $\log_3 x = 2 \Leftrightarrow x = 3^2$.

I. $\log_{0.2} (\log_{0.3} x) = 2 \Leftrightarrow \log_{0.3} x = 0.2^2 = 0.04 \Leftrightarrow x = 0.3^{0.04}$.

J. $\log_{12} (\log_3 x^2) = 3 \Leftrightarrow \log_3 x^2 = 12^3$.

And by the definition again, we can get: $\log_3 x^2 = 12^3 \Leftrightarrow x^2 = 3^{12^3}$.

And we have: $3^{12^3} = 3^{2 \cdot \frac{1}{2} \cdot 12^3} = (3^{\frac{12^3}{2}})^2$.   Thus, we get: $x = 3^{\frac{12^3}{2}}$.

K. $\log_3 (\log_{0.2} x^2) = -2 \Leftrightarrow \log_{0.2} x^2 = 3^{-2} = 1/9$.

And by the definition again, we can get: $\log_{0.2} x^2 = 1/9 \Leftrightarrow x^2 = 0.2^{\frac{1}{9}} = 0.2^{\frac{2}{18}} = (0.2^{\frac{1}{18}})^2$.

Thus, we get: $x = 0.2^{\frac{1}{18}}$.

L. $\log_{0.2} (\log_{12} x^{-2}) = 3 \Leftrightarrow \log_{12} x^{-2} = 0.2^3 = 0.008$.   So we get: $\log_{12} x^{-2} = 0.008$.

And by the definition, we get: $\log_{12} x^{-2} = 0.008 \Leftrightarrow x^{-2} = 12^{0.008} = 12^{\frac{0.008}{-2} \cdot (-2)} = (12^{\frac{0.008}{-2}})^{-2}$.

Thus, we get: $x = 12^{-0.004}$.

M. $\log_2(\log_x 16) = 3 \Leftrightarrow \log_x 16 = 2^3 = 8$.

And by the definition again, we can get: $\log_x 16 = 8 \Leftrightarrow 16 = x^8$.

And we have: $16 = 2^4 = (2^{0.5})^8$.   So we get: $x = 2^{0.5} = \sqrt{2}$.

N. $\log_2(\log_{\sqrt{x}} 8) = -3 \Leftrightarrow \log_{\sqrt{x}} 8 = 2^{-3} = \frac{1}{8}$.

And by the definition again, we can get: $\log_{\sqrt{x}} 8 = \frac{1}{8} \Leftrightarrow 8 = (\sqrt{x})^{\frac{1}{8}} = x^{\frac{1}{16}}$.

And we have: $8 = 2^3 = 2^{3 \cdot 16 \cdot \frac{1}{16}} = 2^{48 \cdot \frac{1}{16}} = (2^{48})^{\frac{1}{16}}$.   So we get: $x = 2^{48}$.

O. $\log_2(\log_{x^2} 3\sqrt{3}) = -2 \Leftrightarrow \log_{x^2} 3\sqrt{3} = 2^{-2} = \frac{1}{4}$.

And by the definition again, we can get: $\log_{x^2} 3\sqrt{3} = \frac{1}{4} \Leftrightarrow 3\sqrt{3} = (x^2)^{\frac{1}{4}} = x^{\frac{1}{2}}$.

And we have: $3\sqrt{3} = 3^{\frac{3}{2}} = (3^3)^{\frac{1}{2}}$.   So we get: $x = 3^3$.

P. $\log_2(\log_{x^{0.3}} 4) = -1 \Leftrightarrow \log_{x^{0.3}} 4 = 2^{-1} = \frac{1}{2} \Leftrightarrow 4 = (x^{0.3})^{\frac{1}{2}} = x^{\frac{0.3}{2}} = x^{\frac{3}{20}}$.

And we have: $4 = 2^2 = 2^{2 \cdot \frac{20}{3} \cdot \frac{3}{20}} = 2^{\frac{40}{3} \cdot \frac{3}{20}} = (2^{\frac{40}{3}})^{\frac{3}{20}}$.   So we get: $x = 2^{\frac{40}{3}}$.

Q. $\log_2\{\log_3(\log_5 x)\} = 2 \Leftrightarrow \log_3(\log_5 x) = 2^2 = 4 \Leftrightarrow \log_5 x = 3^4 = 81 \Leftrightarrow x = 5^{81}$.

Examples 3 on Definitions for Logarithms

Find the value of x in each case below:

0. $x = \log_9 \left(\frac{1}{\sqrt{a}} + \frac{1}{\sqrt{b}} \right)$ where $a + b = 21$ and $ab = 9$.

1. $x = \log_9 \left| \frac{1}{\sqrt{a}} - \frac{1}{\sqrt{b}} \right|$ where $a + b = 9$ and $ab = 9$.

2. $x = \log_{16} \left(\frac{3b}{a^2 + 2} + \frac{3a}{b^2 + 2} \right)$ where $a + b = 6$ and $ab = 2$.

3. $x = \log_{(3\sqrt{2} - 2\sqrt{3})} (a^2 - 4a + 1)$ where $a = \frac{\sqrt{3} + \sqrt{2}}{\sqrt{3} - \sqrt{2}}$.

4. $x = \log_5 (a^2 - ab + b^2)$ where $a = \sqrt{3} - \sqrt{2}$, and $b = \sqrt{3} + \sqrt{2}$.

5. Assuming $\dfrac{1}{\log_x y} + \log_x y = -\dfrac{10}{3}$, find the value of $(x^3 y - 1)(y^3 x - 1)$.

6. Show that $\log_3 2$ is not a rational number.

Suggestions or Solutions
To the Examples 3 on Definitions for Logarithms

The definition for logs is: $A = b^x \Leftrightarrow x = \log_b A$.

So using the definition, we can get the solutions the way below.

0. $x = \log_9(\frac{1}{\sqrt{a}} + \frac{1}{\sqrt{b}})$ **where** $a + b = 21$ **and** $ab = 9$.

First, we can have: $\frac{1}{\sqrt{a}} + \frac{1}{\sqrt{b}} = \frac{\sqrt{a}+\sqrt{b}}{\sqrt{ab}}$, which is positive.

And next, squaring it, we get: $(\frac{\sqrt{a}+\sqrt{b}}{\sqrt{ab}})^2 = \frac{a+b+2\sqrt{ab}}{ab} = \frac{21+2\sqrt{9}}{9} = \frac{27}{9} = 3$.

So we can get: $\frac{\sqrt{a}+\sqrt{b}}{\sqrt{ab}} = \sqrt{3}$. And we know: $\frac{1}{\sqrt{a}} + \frac{1}{\sqrt{b}} = \frac{\sqrt{a}+\sqrt{b}}{\sqrt{ab}}$.

Thus, we get: $x = \log_9(\frac{1}{\sqrt{a}} + \frac{1}{\sqrt{b}}) = \log_9 \sqrt{3}$. So we get: $x = \log_9 \sqrt{3}$.

And by the definition, we get: $x = \log_9 \sqrt{3} \Leftrightarrow \sqrt{3} = 9^x$.

And we have: $\sqrt{3} = 3^{0.5}$, and $9^x = 3^{2x}$. So we get: $\sqrt{3} = 9^x \Rightarrow 3^{0.5} = 3^{2x}$.

Thus, we get: $2x = 0.5 \Rightarrow x = 1/4$.

In short:

To begin with, we can get: $\frac{1}{\sqrt{a}} + \frac{1}{\sqrt{b}} = \frac{\sqrt{a}+\sqrt{b}}{\sqrt{ab}}$, and $(\frac{\sqrt{a}+\sqrt{b}}{\sqrt{ab}})^2 = \frac{a+b+2\sqrt{ab}}{ab} = \frac{21+2\sqrt{9}}{9} = \frac{27}{9} = 3$.

So we can get: $\frac{\sqrt{a}+\sqrt{b}}{\sqrt{ab}} = \sqrt{3} \Rightarrow x = \log_9(\frac{1}{\sqrt{a}} + \frac{1}{\sqrt{b}}) = \log_9 \sqrt{3} \Rightarrow x = \log_9 \sqrt{3} \Leftrightarrow \sqrt{3} = 9^x$.

And thus, we get: $\sqrt{3} = 9^x \Rightarrow 3^{0.5} = 3^{2x} \Rightarrow 2x = 0.5 \Rightarrow x = 1/4$.

1. $x = \log_9 \left| \frac{1}{\sqrt{a}} - \frac{1}{\sqrt{b}} \right|$ **where** $a + b = 9$ **and** $ab = 9$.

First, we can have: $\frac{1}{\sqrt{a}} - \frac{1}{\sqrt{b}} = \frac{\sqrt{a} - \sqrt{b}}{\sqrt{ab}}$, of which we don't know the sign.

And next, squaring it, we get: $\left(\frac{\sqrt{a} - \sqrt{b}}{\sqrt{ab}} \right)^2 = \frac{a + b - 2\sqrt{ab}}{ab} = \frac{9 - 2\sqrt{9}}{9} = \frac{3}{9} = \frac{1}{3}$.

And we know: $\left| \frac{\sqrt{a} - \sqrt{b}}{\sqrt{ab}} \right| > 0$ if $a \neq b$, of course.

So we can get: $\left| \frac{\sqrt{a} - \sqrt{b}}{\sqrt{ab}} \right| = \sqrt{\frac{1}{3}}$. And we know: $\left| \frac{1}{\sqrt{a}} - \frac{1}{\sqrt{b}} \right| = \left| \frac{\sqrt{a} - \sqrt{b}}{\sqrt{ab}} \right|$.

Thus, we get: $x = \log_9 \left| \frac{1}{\sqrt{a}} - \frac{1}{\sqrt{b}} \right| = \log_9 \sqrt{\frac{1}{3}}$. So we get: $x = \log_9 \sqrt{\frac{1}{3}}$.

And by the definition, we get: $x = \log_9 \sqrt{\frac{1}{3}} \Leftrightarrow \sqrt{\frac{1}{3}} = 9^x$.

And we have: $\sqrt{\frac{1}{3}} = \frac{1}{\sqrt{3}} = \frac{1}{3^{0.5}} = 3^{-0.5}$, and $9^x = 3^{2x}$. So we get: $\sqrt{\frac{1}{3}} = 9^x \Rightarrow 3^{-0.5} = 3^{2x}$.

Thus, we get: $2x = -0.5 \Rightarrow x = -1/4$.

In short:

To begin with, we can get: $\frac{1}{\sqrt{a}} - \frac{1}{\sqrt{b}} = \frac{\sqrt{a} - \sqrt{b}}{\sqrt{ab}}$, and $\left(\frac{\sqrt{a} - \sqrt{b}}{\sqrt{ab}} \right)^2 = \frac{a + b - 2\sqrt{ab}}{ab} = \frac{9 - 2\sqrt{9}}{9} = \frac{3}{9} = \frac{1}{3}$.

So we can get: $\left| \frac{\sqrt{a} - \sqrt{b}}{\sqrt{ab}} \right| = \sqrt{\frac{1}{3}} \Rightarrow x = \log_9 \left| \frac{1}{\sqrt{a}} - \frac{1}{\sqrt{b}} \right| = \log_9 \sqrt{\frac{1}{3}} \Rightarrow x = \log_9 \sqrt{\frac{1}{3}} \Leftrightarrow \sqrt{\frac{1}{3}} = 9^x$.

And thus, we get: $\sqrt{\frac{1}{3}} = 9^x \Rightarrow 3^{-0.5} = 3^{2x} \Rightarrow 2x = -0.5 \Rightarrow x = -1/4$.

2. $x = \log_{16}\left(\frac{3b}{a^2+2} + \frac{3a}{b^2+2}\right)$ **where** $a + b = 6$ **and** $ab = 2$.

To begin with, we can have: $\frac{3b}{a^2+2} + \frac{3a}{b^2+2} = \frac{3b(b^2+2)+3a(a^2+2)}{(a^2+2)(b^2+2)} = \frac{3b^3+3a^3+6(a+b)}{a^2b^2+2(a^2+b^2)+4} = \frac{3(b^3+a^3)+6(a+b)}{a^2b^2+2(a^2+b^2)+4}$.

Next, we can have: $a^2 + b^2 = (a + b)^2 - 2ab$.

So we get: $a^2 + b^2 = 6^2 - 2\cdot 2 = 32$.

And next, we can have: $a^3 + b^3 = (a + b)^3 - 3ab(a + b)$.

So we get: $a^3 + b^3 = 6^3 - 3\cdot 2\cdot 6 = 6^3 - 6^2 = 6^2(6 - 1) = 36\cdot 5 = 180$.

Thus, we get: $\frac{3(b^3+a^3)+6(a+b)}{a^2b^2+2(a^2+b^2)+4} = \frac{3\cdot 180+6\cdot 6}{2^2+2\cdot 32+4} = \frac{540+36}{4+64+4} = \frac{576}{72} = 8$.

So we get: $x = \log_{16}\left(\frac{3b}{a^2+2} + \frac{3a}{b^2+2}\right) = \log_{16} 8 \Rightarrow x = \log_{16} 8$.

And by the definition, we get: $x = \log_{16} 8 \Leftrightarrow 8 = 16^x$.

And we have: $8 = 2^3$, and $16^x = 2^{4x}$.

Thus, we get: $4x = 3 \Rightarrow x = 3/4$.

3. $x = \log_{(3\sqrt{2}-2\sqrt{3})}(a^2 - 4a + 1)$ **where** $a = \frac{\sqrt{3}+\sqrt{2}}{\sqrt{3}-\sqrt{2}}$.

To begin with, we can have: $a = \frac{\sqrt{3}+\sqrt{2}}{\sqrt{3}-\sqrt{2}} = \frac{(\sqrt{3}+\sqrt{2})^2}{(\sqrt{3}-\sqrt{2})(\sqrt{3}+\sqrt{2})} = \frac{3+2+2\sqrt{6}}{3-2} = 5 + 2\sqrt{6}$.

So next, we can have: $a^2 - 4a + 1 = (5+2\sqrt{6})^2 - 4(5+2\sqrt{6}) + 1$

$= 25 + 24 + 20\sqrt{6} - 20 - 8\sqrt{6} + 1 = 30 - 12\sqrt{6} = 2(15 - 6\sqrt{6}) = 2(15 - 2\sqrt{9\cdot6})$

$= 2(9 + 6 - 2\sqrt{9\cdot6}) = 2(\sqrt{9} - \sqrt{6})^2 = 2(3 - \sqrt{6})^2$

$= (\sqrt{2})^2(3 - \sqrt{6})^2 = (3\sqrt{2} - \sqrt{12})^2 = (3\sqrt{2} - 2\sqrt{3})^2$.

So we get: $x = \log_{(3\sqrt{2}-2\sqrt{3})}(a^2 - 4a + 1) = \log_{(3\sqrt{2}-2\sqrt{3})}(3\sqrt{2} - 2\sqrt{3})^2$.

And next, setting: $b = 3\sqrt{2} - 2\sqrt{3}$, we get: $x = \log_b b^2$.

And by the definition, we get: $x = \log_b b^2 \Leftrightarrow b^2 = b^x$. Thus, we get: $x = 2$.

4. $x = \log_5(a^2 - ab + b^2)$ **where** $a = \sqrt{3} - \sqrt{2}$, and $b = \sqrt{3} + \sqrt{2}$.

To begin with, we can have: $a^2 - ab + b^2 = (a - b)^2 + ab$.

So next, we can have: $a^2 - ab + b^2 = \{(\sqrt{3} - \sqrt{2}) - (\sqrt{3} + \sqrt{2})\}^2 + (\sqrt{3} - \sqrt{2})(\sqrt{3} + \sqrt{2})$

$= (-\sqrt{2})^2 + (3 - 2) = 4 + 1 = 5$.

So we get: $x = \log_5(a^2 - ab + b^2) = \log_5 5 \Rightarrow x = \log_5 5$.

And by the definition, we get: $x = \log_5 5 \Leftrightarrow 5 = 5^x$. Thus, we get: $x = 1$.

5. Assuming $\dfrac{1}{\log_x y} + \log_x y = -\dfrac{10}{3}$, find the value of $(x^3 y - 1)(y^3 x - 1)$.

Assuming first, $a = \log_x y$, we can set: $1/a + a = -10/3$.

Next, we can solve for a the equation above the way below:

$1/a + a = -10/3 \Rightarrow 1 + a^2 = -10a/3 \Rightarrow 3 + 3a^2 = -10a \Rightarrow 3a^2 + 10a + 3 = 0$.

$3a^2 + 10a + 3 = 3a^2 + 9a + a + 3 = 3a(a + 3) + (a + 3) = (a + 3)(3a + 1) = 0$.

So we get: $a = -3$ or $-1/3$. And we know: $a = \log_x y$.

So first, we can get: $\log_x y = -3 \Rightarrow y = x^{-3}$ by the definition for logs.
And next, by the definition for logs again, we can get: $\log_x y = -1/3 \Rightarrow y = x^{-1/3} \Rightarrow x = y^{-3}$.

So putting threads together, we have: $y = x^{-3}$ or $x = y^{-3}$.
And we can put them this way, too: $x^3 y = 1$ or $y^3 x = 1$.

So first, $x^3 y = 1 \Rightarrow (x^3 y - 1)(y^3 x - 1) = 0$.
And next, $y^3 x = 1 \Rightarrow (x^3 y - 1)(y^3 x - 1) = 0$.

6.   Show that $\log_3 2$ is not a rational number.

Assuming $\log_3 2$ is a rational number, we can set: $\log_3 2 = m/n$ where m and n are integers prime to each other.

Then, by the definition for logs, we get: $2 = 3^{m/n}$. So we get: $2^n = 3^m$, which is not possible, because 2^n is even, but 3^m is odd.

So it is not the case where $2^n = 3^m$, which means it cannot be the case where we can set $\log_3 2 = m/n$ where m and n are integers prime to each other, which means $\log_3 2$ is not a rational number.

8. **Bases Matter in Logarithms**

To begin with, what is a logarithm?

It is called a log for short, and is no other than an exponent, which is a number we use to make a power as 2^3, where 2 is called the base, and 3 is called the exponent.

And taking the log of a power to the base used in the power, we get the exponent used in the power.
So for instance, taking the log of 2^3 to the base 2, we get: $\log_2 2^3$, which is the exponent used in the power 2^3. So $\log_2 2^3$ is the exponent 3. Thus, we get: $3 = \log_2 2^3$.

And we know: $2^3 = 8$, so we get: $\log_2 2^3 = \log_2 8 = 3$.

And thus, we can put all the ideas above at once the way below:

$$8 = 2^3 \Leftrightarrow 3 = \log_2 8 \text{ or } 3 = \log_2 2^3.$$

In fact, the definition for logs is: $A = b^x \Leftrightarrow x = \log_b A$, where A and $b > 0$, but $b \neq 1$.

And since we have: $A = b^x$, we can get this, too: $x = \log_b b^x$.

So using the *definition for logs*, we can *switch between a power and its exponent*.
We don't just switch between the two, though. How then, do we do so?

We do so based on the base. What base though?

It is the base we use taking a log of a number, and we use it making a power, too.
We can make many different powers that have the same value. Those powers then, have different bases. For instance:

$$16 = 16^1 = 4^2 = 2^4 = (\sqrt{2})^8 = (\sqrt[3]{2})^{12} = \ldots$$

What then, do we get taking a log of 16?

It depends on the base to which we take a log of 16.

Taking the log of 16 to base 4, we get 2, which is the exponent in the power 4^2.

Taking the log of 16 to base 2, we get 4, which is the exponent in the power 2^4.

Taking the log of 16 to base $\sqrt{2}$, we get 8, which is the exponent in the power $(\sqrt{2})^8$.

And so forth

Taking thus, a log of the same number to a different base, we get a different log, that is, a different exponent.

So the base matters.

And we can get:

$16^1 \Leftrightarrow 1 = \log_{16} 16$ $\qquad 4^2 \Leftrightarrow 2 = \log_4 16$ $\qquad 2^4 \Leftrightarrow 4 = \log_2 16$

$(\sqrt{2})^8 \Leftrightarrow 8 = \log_{\sqrt{2}} 16$ $\qquad (\sqrt[3]{2})^{12} \Leftrightarrow 12 = \log_{\sqrt[3]{2}} 16$ $\quad \ldots$

And thus, the base can be said to be the pivot of switching between the two, a power and its exponent.
And we can also say that using the *definition for logs*, we can switch between an antilog and its log, too. And the switching is made based on the base.

And we have another case where the base matters, too. So let's see now, what the case is.

Assuming first, *b* is the base of a power, and the exponent is *x*, we can put the power the way as follows: b^x.

Then, if changing the value of the base b keeping constant the value of the power b^x, we have to change the value of the exponent x, too.

In short, changing b holding b^x constant, we have to change x, too.
What then, about a logarithm?

For instance, taking a log of a number A to a base b, we get: $\log_b A$.
And we know that a log is an exponent. So assuming x is the exponent to which the base b is raised to equal the number A, we get: $x = \log_b A$. Then, we call A the antilog.

And we have this, too: $A = b^x$, since A is the number we get raising the base b to the exponent x.
And also, we can get it by the definition, too, which is: $A = b^x \Leftrightarrow x = \log_b A$, where A and $b > 0$, yet $b \neq 1$.

Now, keeping A constant, if we change the base b, the exponent x changes, too.

So keeping constant the antilog in a log, and using a different number as the base, we get a different value of the log.
And thus, when we take the log of a number to a base, the value of the log depends on the value of the base. In short, a log depends on the base we choose.

So the base matters in a log.

Depending on the extent of the base, a log works two different ways, which are opposite of each other. So the base in a log can be in one of two different cases.

In one, the <u>base is between 0 and 1</u>, and in the other, the <u>base is greater than 1</u>.

And the same is true for powers, too, of course.

Thus, doing problems on powers and logarithms, we want to consider the two cases as follows: "**0 < the base < 1**", and "**the base > 1**"

So let's now take a look at how powers and logs work in each case.

Suppose first, that in $A = b^x$, the exponent $x \geq 0$, and that the base b is between 0 and 1. In short, $x \geq 0$, and $0 < b < 1$ in $A = b^x$. Then, we get: $0 < A \leq 1$.   How come?

Assuming first, $c > 1$, we can get: $\frac{1}{c} > 0$ as well as $\frac{1}{c} < 1$.

So we get: $0 < \frac{1}{c} < 1$.   How come though, we get: $\frac{1}{c} > 0$, too?

Since $c > 1$, as c increases, $\frac{1}{c}$ decreases, but stays positive, and also, cannot be 0 no matter how large c may be. So we get: $\frac{1}{c} > 0$, together with $\frac{1}{c} < 1$. And thus, $0 < \frac{1}{c} < 1$.

So assuming next, $b = \frac{1}{c}$, we get: $0 < b < 1$.

Now, we have: $A = b^x$, where $b = \frac{1}{c}$.   So we get: $A = (\frac{1}{c})^x = \frac{1}{c^x}$, and thus, we get: $A = \frac{1}{c^x}$.

Meanwhile, $x \geq 0 \Rightarrow c^x \geq 1$, because $c > 1$, and if $x = 0$, we get: $c^x = c^0 = 1$.

So we can get: not only $A = \frac{1}{c^x} \leq 1$ but $A = \frac{1}{c^x} > 0$, too.

That's because since $c^x \geq 1$, as c^x increases, $\frac{1}{c^x}$ decreases, but stays positive, and cannot be 0 no matter how large c^x may be.   So we get: $0 < \frac{1}{c^x} \leq 1$.

And we know: $A = \frac{1}{c^x}$.   So we get: $0 < A \leq 1$.

Now, we have set: $A = b^x$, where $b = \frac{1}{c}$ for $c > 1$.

So we now have: $A = \frac{1}{c^x} = b^x$, where $0 < b < 1$.

Therefore, if $x \geq 0$ when $0 < b < 1$, we get: $0 < A \leq 1$.

Let's see now, more specifically, how the value of $A = b^x$ changes as that of x changes.

We are now considering the case where $0 < b < 1$.

So suppose for instance, $b = 0.5 = \frac{1}{2}$. Then:

If $x = 0$, we get $A = \left(\frac{1}{2}\right)^0 = 1$.

If $x = \frac{1}{2}$, we get $A = \left(\frac{1}{2}\right)^{\frac{1}{2}} = \sqrt{\frac{1}{2}} = \frac{1}{\sqrt{2}} = \frac{\sqrt{2}}{2}$, which is between 0 and 1, since $\sqrt{2} < 2$.

If $x = 1$, we get $A = \left(\frac{1}{2}\right)^1 = \frac{1}{2}$.

If $p = 3$, we get $A = \left(\frac{1}{2}\right)^3 = \frac{1}{8}$. ... ...

If $x = 1000$, we get $A = \left(\frac{1}{2}\right)^{1000} = \frac{1}{2^{1000}}$, which is very close to 0.

So we can see that if the value of the exponent x begins with 0, and gets larger, the value of the power b^x begins with 1, but gets smaller, and approaches 0, but will never be 0.

Let's now, put the ideas above in log's perspective.

To begin with, the short definition for logs is as follows: $A = b^x \Leftrightarrow x = \log_b A$.

So x is the value of $\log_b A$ as well as the exponent in the power b^x.

Referring thus, to the values of x and A listed in the case where $b = 0.5$, we can say that when the base b is between 0 and 1, if the value of the antilog A begins with 1, decreases, and approaches 0, the value of x, that is, the value of $\log_b A$ begins with 0, and keeps increasing.

And in particular, the change of x is far greater than the change of A.

Let's now, put the ideas above in a graph.

44

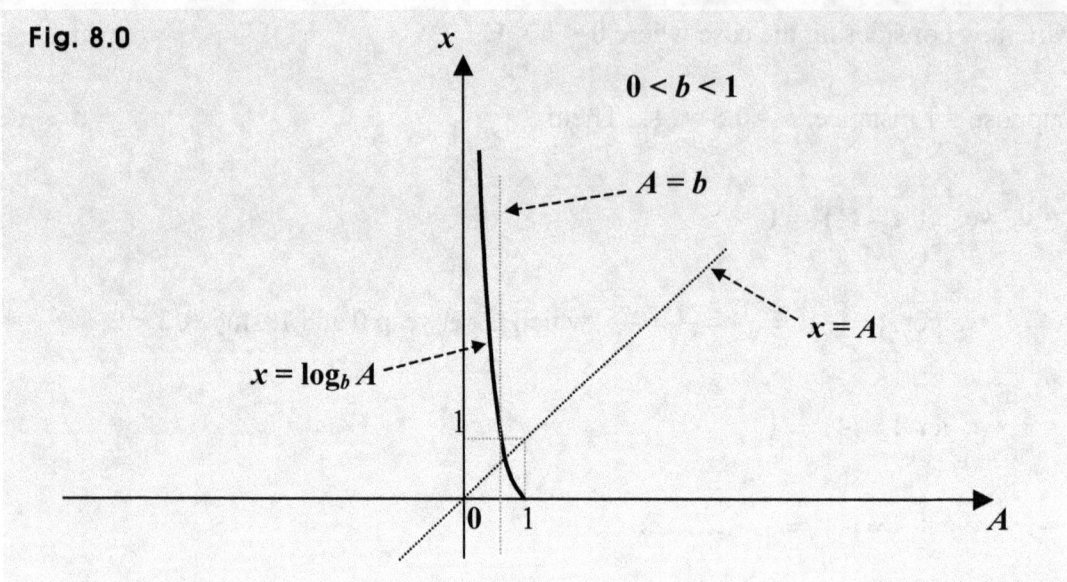

Fig. 8.0

$0 < b < 1$

$A = b$

$x = A$

$x = \log_b A$

Suppose next, the exponent $x \leq 0$, and $0 < b < 1$ in $A = b^x$.

Then, we get: $A \geq 1$.   How come?

Suppose again, that $b = \frac{1}{c}$ where $c > 1$.

Then, we get: $0 < \frac{1}{c} < 1$, and thus, we get: $0 < b < 1$.

Now, we have: $A = b^x$, where $b = \frac{1}{c}$. So $A = (\frac{1}{c})^x = c^{-x}$, and thus, we get: $A = c^{-x}$.

Meanwhile, $x \leq 0 \Rightarrow -x \geq 0 \Rightarrow c^{-x} \geq 1$, because $c > 1$, and if $-x = 0$, we get: $c^{-x} = c^0 = 1$.

And we have: $A = c^{-x} = b^x$.

So if $x \leq 0$ when $0 < b < 1$, we get: $A \geq 1$.

Let's see now, more specifically, how the value of $A = b^x$ changes as that of x changes.

Suppose again, that $b = 0.5 = \frac{1}{2}$. Then:

If $x = 0$, we get: $A = \left(\frac{1}{2}\right)^0 = 1$.

If $x = -\frac{1}{4}$, we get: $A = \left(\frac{1}{2}\right)^{-\frac{1}{4}} = 2^{\frac{1}{4}} = \sqrt[4]{2} > 1$. How come we get: $\sqrt[4]{2} > 1$?

Using a calculator, of course, we can readily see that: $\sqrt[4]{2} > 1.189$.

Let's check algebraically though, to see if $\sqrt[4]{2} > 1$.

To begin with, we can readily see that if $m > 1$, we get: $m^4 > 1$, since $1^4 = 1$, and $2^4 = 16$.

If however, we are given $(m^4 > 1)$ only, then can we just say that: $m > 1$, too?

No, we can't, because m can be negative, too. For instance, $(-2)^4 = 16 > 1$ but $-2 < 1$.

If we have though, $m > 0$ and $m^4 > 1$, we can say that we get: $m > 1$.

Now, we know: $\sqrt[4]{2} > 0$, and we have: $(\sqrt[4]{2})^4 = 2 > 1$, so we get: $\sqrt[4]{2} > 1$.

Now, getting back to the point where we left off, we have: $b = 0.5 = \frac{1}{2}$, and have taken two steps as follows:

If $x = 0$, we get: $A = \left(\frac{1}{2}\right)^0 = 1$. And if $x = -\frac{1}{4}$, we get: $A = \sqrt[4]{2}$.

Let's now, take several more steps.

If $x = -\frac{1}{2}$, we get: $A = \left(\frac{1}{2}\right)^{-\frac{1}{2}} = 2^{\frac{1}{2}} = \sqrt{2}$.

If $x = -1$, we get: $A = \left(\frac{1}{2}\right)^{-1} = 2$.

If $x = -3$, we get: $A = \left(\frac{1}{2}\right)^{-3} = 2^3 = 8$.

…

If $x = -10$, we get: $A = \left(\frac{1}{2}\right)^{-10} = 2^{10} = 1024$.

Thus, if the value of the exponent x starts at 0, and gets smaller, the value of the power b^x begins with 1, and gets larger and larger.

Let's now, put the ideas above in log's perspective.

First, beginning with the short definition for logs, we have: $A = b^x \Leftrightarrow x = \log_b A$.

So x is the value of $\log_b A$ as well as the exponent in the power b^x.

Therefore, referring to the values of x and A in the case where $b = 0.5$, and $x < 0$, we can say that when the base b is between 0 and 1, if the value of the antilog A begins with 1, and increases, the value of x, i.e., the value of $\log_b A$ begins with 0, and keeps decreasing.

This time however, the change of x is much smaller than the change of A.

Let's now, put in a graph the ideas above, together with the curve in the **Fig. 8.0**.

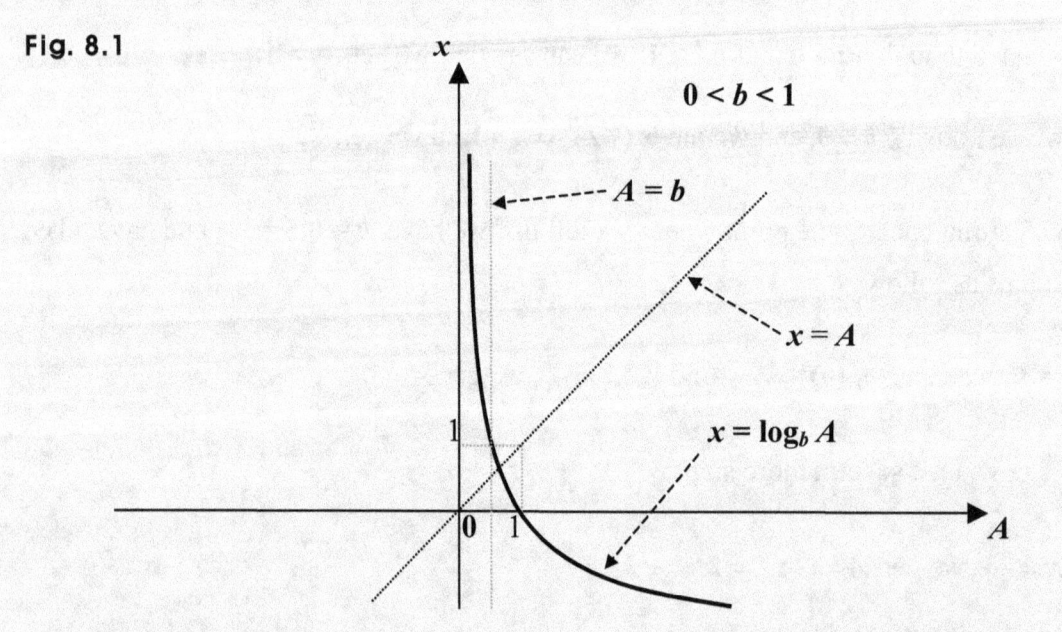

Fig. 8.1

$0 < b < 1$

$A = b$

$x = A$

$x = \log_b A$

Therefore, we can see that if $0 < Y \le X$, we get: $\log_b Y \ge \log_b X$ where $0 < b < 1$.

That is to say that if the base is between 0 and 1, the larger the antilog is, the smaller the log gets.

Let's now, put together the two cases where $x \geq 0$ and $x \leq 0$, and then, reconsider both. This time though, we will put the steps in reverse order.

We have: $b = \frac{1}{2}$, and $A = b^x \Leftrightarrow x = \log_b A$. Then:

If $x = -10$, we get: $A = (\frac{1}{2})^{-10} = 2^{10} = 1024$.

... ...

If $x = -3$, we get: $A = (\frac{1}{2})^{-3} = 2^3 = 8$.

If $x = -1$, we get: $A = (\frac{1}{2})^{-1} = 2$.

If $x = -\frac{1}{2}$, we get: $A = (\frac{1}{2})^{-\frac{1}{2}} = 2^{\frac{1}{2}} = \sqrt{2}$.

If $x = -\frac{1}{4}$, we get: $A = \sqrt[4]{2}$.

If $x = 0$, we get: $A = (\frac{1}{2})^0 = 1$.

If $x = \frac{1}{2}$, we get $A = (\frac{1}{2})^{\frac{1}{2}} = \sqrt{\frac{1}{2}} = \frac{1}{\sqrt{2}} = \frac{\sqrt{2}}{2}$.

If $x = 1$, we get $A = (\frac{1}{2})^1 = \frac{1}{2}$.

If $x = 3$, we get $A = (\frac{1}{2})^3 = \frac{1}{8}$.

... ...

If $x = 1000$, we get: $A = (\frac{1}{2})^{1000} = \frac{1}{2^{1000}}$, which is very close to 0.

So we can see that as the exponent x gets larger, the antilog A, that is, the value of the power b^x gets smaller, and approaches 0, but will never be 0.

Let's now, put in a graph the ideas above.

Fig. 8.2

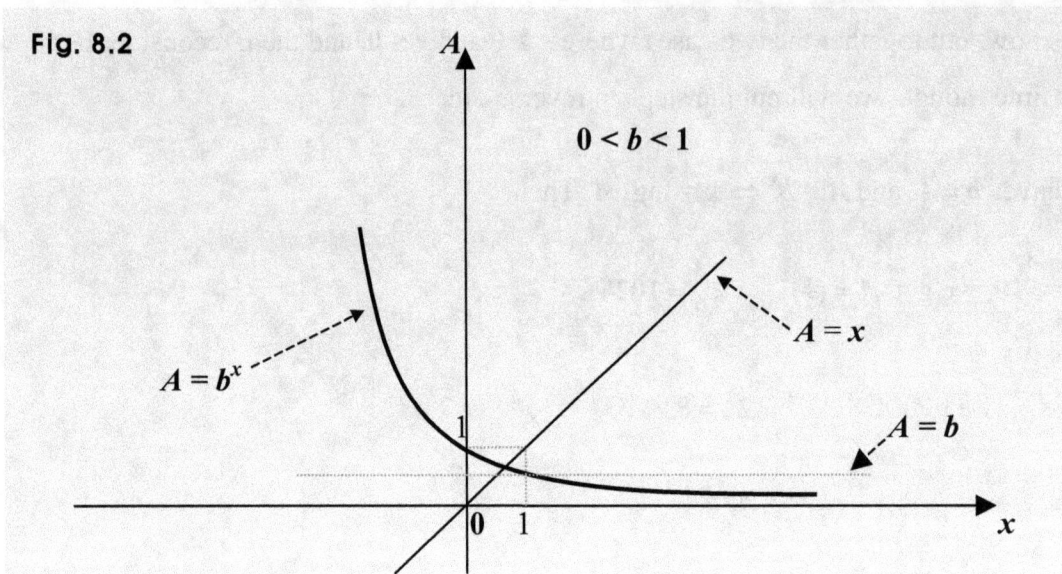

In the graph above, we can see that if $Y \leq X$, then we get: $b^Y \geq b^X$ where $0 < b < 1$.

And putting together the curve in the graph above and the one in the **Fig. 8.1**, we can put the two curves in one graph as shown below.

Note that the two curves share the same variables since we put the two in one graph.

Each value of x in $y = b^x$ is an exponent, but in $y = \log_b x$, each value of x is an antilog.

Fig. 8.3

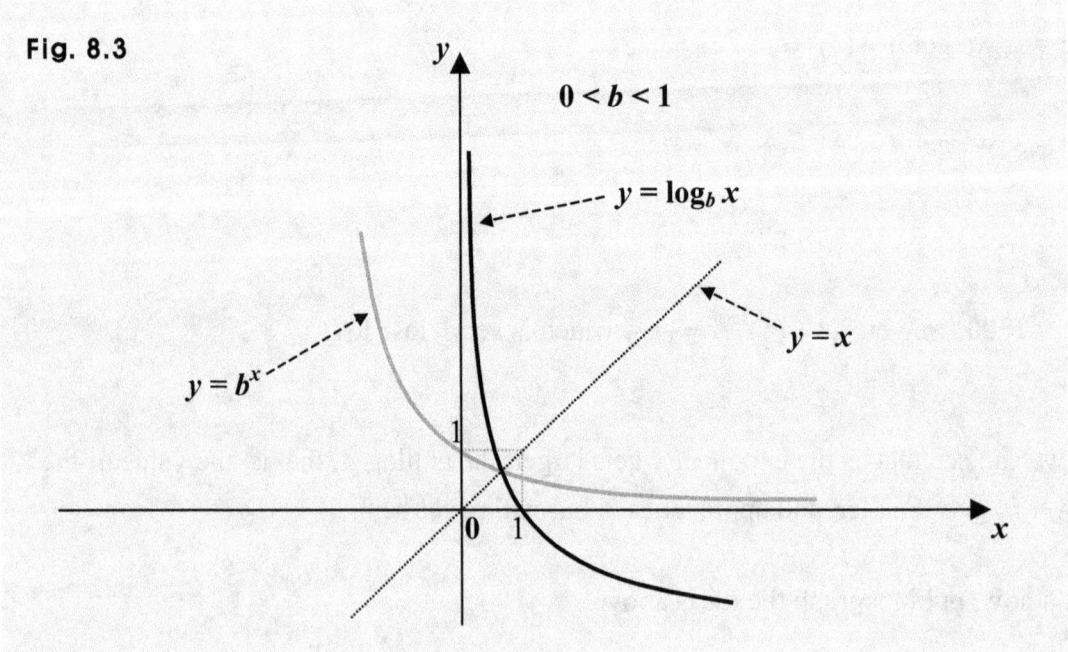

We can notice that there is symmetry between the two curves. The axis of symmetry is a line $y = x$.   How come?

Refer to **Basic Functions** or **Graph Operations**.

Suppose next, the exponent $x \geq 0$, and the base $b > 1$ in $A = b^x$.

Then, we get: $A \geq 1$.   How come?

Assuming $b > 1$, we get: $x \geq 0 \Rightarrow b^x \geq 1$, because $b > 1$ and if $x = 0$, we get: $b^x = b^0 = 1$.

And we have $A = b^x$.

Therefore, if $x \geq 0$ when $b > 1$, we get: $A \geq 1$. Let's see now, how the value of A changes as the value of x changes.

Suppose for instance, $b = 2$. Then:

If $x = 0$, we get: $A = 2^0 = 1$.

If $x = \frac{1}{2}$, we get $A = \sqrt{2}$.

If $x = 1$, we get: $A = 2$.

If $x = 3$, we get: $A = 2^3 = 8$.

...

If $x = 10$, we get: $A = 2^{10} = 1024$.

So we can see that if the value of the exponent x begins with 0, and gets larger, the value of the antilog A begins with 1, and gets larger.

And we can put it this way, too: if the value of the antilog A begins with 1, and gets larger, the value of x begins with 0, and gets larger, that is, the value of $\log_b A$ gets larger.

Let's now, put the ideas in a graph.

50

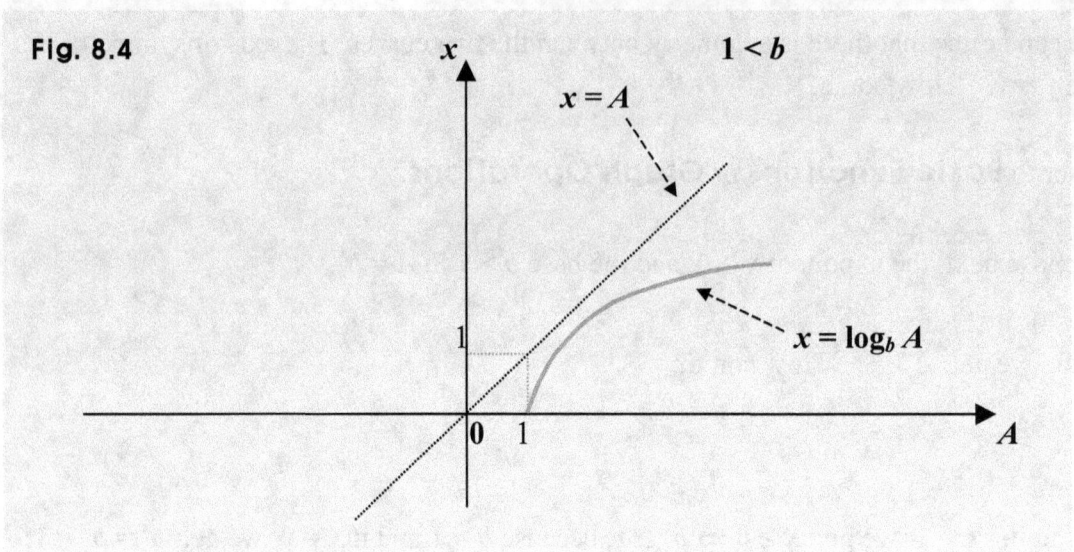

Fig. 8.4 $1 < b$

Suppose next, the exponent $x \le 0$, and the base $b > 1$ in $A = b^x$.

Then, we get: $0 < A \le 1$. How come?

We have $A = b^x$.   So $A = (b^{-1})^{-x} = (\frac{1}{b})^{-x}$, and thus, we get: $A = (\frac{1}{b})^{-x}$.

And we have $b > 1$.   So we get: $\frac{1}{b} < 1$.

Suppose now, that $\frac{1}{b} = c$.   Then, we get: $c < 1$ since we have: $\frac{1}{b} < 1$.

And we get: $0 < c$, too, because as b increases, $c = \frac{1}{b}$ decreases, but stays positive, but cannot be 0 no matter how large b may be.   So we get: $0 < c < 1$.

Let's next, put A this way: $A = c^{-x}$, since we have: $c = \frac{1}{b}$, and $A = (\frac{1}{b})^{-x}$.

Next, $x \le 0 \Rightarrow -x \ge 0 \Rightarrow 0 < c^{-x} \le 1$, because $0 < c < 1$, and if $-x = 0$, we get: $c^{-x} = c^0 = 1$.

So we now have: $A = c^{-x} = (\frac{1}{b})^{-x} = b^x$.

Therefore, if $x \le 0$ when $b > 1$, we get: $0 < A \le 1$.

Let's see now, in $A = b^x$, how the value of A changes as that of x changes.

Suppose again, that $b = 2$. Then:

If $x = 0$, we get: $A = 2^0 = 1$.

If $x = -\frac{1}{4}$, we get: $A = 2^{-\frac{1}{4}} = \sqrt[4]{\frac{1}{2}} = \sqrt[4]{\frac{2^3}{2^4}} = \frac{\sqrt[4]{2^3}}{\sqrt[4]{2^4}} = \frac{\sqrt[4]{2^3}}{2}$.

If $x = -\frac{1}{2}$, we get: $A = 2^{-\frac{1}{2}} = \sqrt{\frac{1}{2}} = \sqrt{\frac{2}{2^2}} = \frac{\sqrt{2}}{\sqrt{2^2}} = \frac{\sqrt{2}}{2}$.

If $x = -1$, we get: $A = 2^{-1} = \frac{1}{2}$.

If $p = -3$, we get: $A = 2^{-3} = \frac{1}{2^3} = \frac{1}{8}$.

... ...

If $x = -1000$, we get: $A = 2^{-1000} = \frac{1}{2^{1000}}$, which is pretty close to 0.

So if the exponent x starts at 0, and gets smaller, the antilog A begins with 1, gets smaller, and approaches 0, but will never be 0.

Then again, considering $A = b^x \Leftrightarrow x = \log_b A$, we can put the idea the way below, too:

If the antilog A starts at 1, gets smaller, and approaches 0, the value of $x = \log_b A$ starts at 0, and gets smaller.

Let's now put in a graph the ideas above, together with the curve in the **Fig. 8.4**.

Fig. 8.5

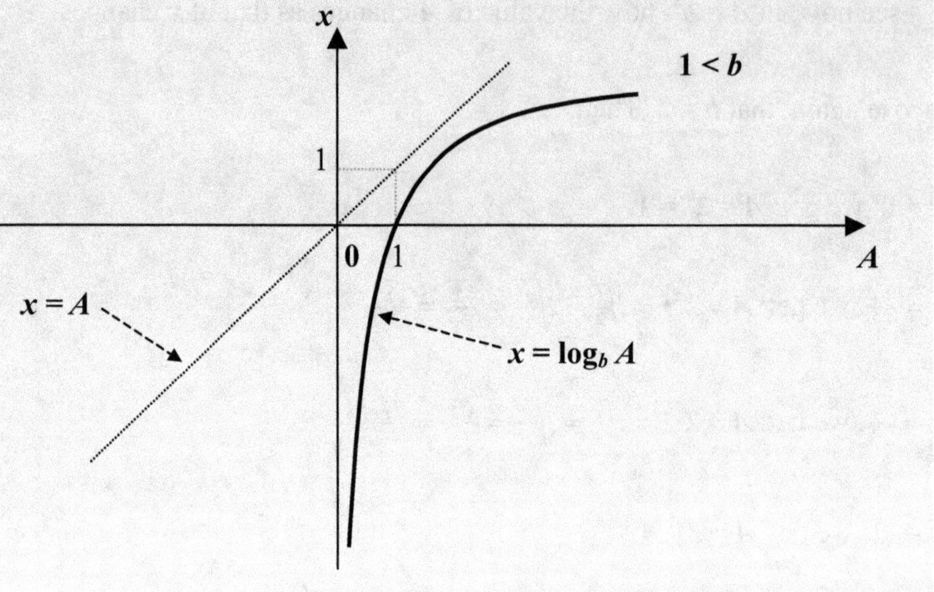

Therefore, we can see that if $0 < Y \le X$, we get: $\log_b Y \le \log_b X$ where $1 < b$.

Let's now, put together the two cases where $x \ge 0$ and $x \le 0$, and then, reconsider both. This time though, we will take the steps backward.

We have $b = 2$, and $A = b^x \Leftrightarrow x = \log_b A$. Then:

If $x = 10$, we get: $A = 2^{10} = 1024.$ ...

If $x = 3$, we get: $A = 2^3 = 8$.   If $x = 1$, we get: $A = 2$.

If $x = \frac{1}{2}$, we get: $A = \sqrt{2}$.   If $x = 0$, we get: $A = 2^0 = 1$.

If $x = -\frac{1}{4}$, we get: $A = 2^{-\frac{1}{4}} = \frac{1}{2^{\frac{1}{4}}} = \frac{2^{\frac{3}{4}}}{2^{\frac{4}{4}}} = \frac{\sqrt[4]{2^3}}{2}$.   If $x = -\frac{1}{2}$, we get: $A = 2^{-\frac{1}{2}} = \frac{1}{2^{\frac{1}{2}}} = \frac{2^{\frac{1}{2}}}{2^{\frac{2}{2}}} = \frac{\sqrt{2}}{2}$.

If $x = -1$, we get: $A = 2^{-1} = \frac{1}{2}$.   If $x = -3$, we get: $A = 2^{-3} = \frac{1}{2^3} = \frac{1}{8}$. ... ...

If $x = -1000$, we get: $A = 2^{-1000} = \frac{1}{2^{1000}}$, which is quite close to 0.

So we can see that if the exponent x begins with 10, gets smaller, and proceeds to negative infinity, the antilog A starts at 1024, gets smaller, and approaches 0, but will never be 0. Let's now, put the ideas in a graph.

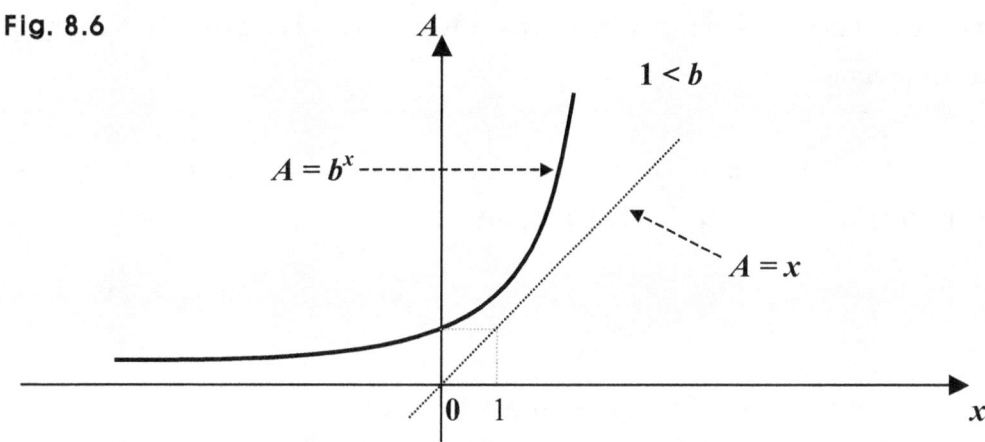

Fig. 8.6

In the graph above, we can see that if $Y \leq X$, we get: $b^Y \leq b^X$ where $1 < b$.

And putting together the curve in the graph above and the one in the **Fig. 8.5**, we get a graph as shown below. Note that the two curves share the same variables since we put the two in one graph. Each value of y in $y = b^x$ is an antilog, but in $y = \log_b x$, each value of y is a log value, that is, an exponent.

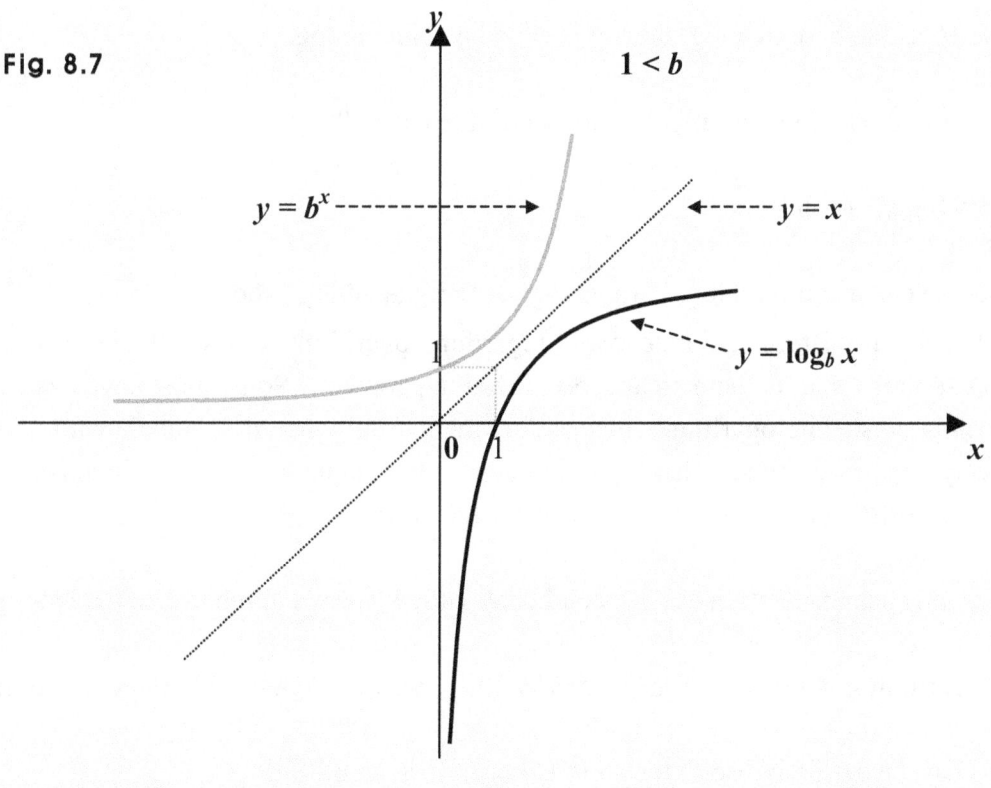

Fig. 8.7

We can notice that there is symmetry between the two curves. The axis of symmetry is a line $y = x$. How come?

That's because the two equations are inverses of each other. If not sure of inverses, refer to **Basic Functions** or **Graph Operations**.

In short:

If we have: $0 < b < 1$ and $X \geq Y$, we get: $\log_b X \leq \log_b Y$.

Besides, if we have: $0 < b < 1$ and $X \geq Y$, we get: $b^X \leq b^Y$.

If we have $b > 1$ and $X \geq Y$, we get: $\log_b X \geq \log_b Y$.

Besides, if we have $b > 1$ and $X \geq Y$, we get $b^X \geq b^Y$.

And there are some categories in logarithms.

If we use 10 as the base in a log, the log is called a *common* log.

In such a case though, we usually do not specify the base 10.

So **log 7 = log₁₀7**, for instance.

Some people call a common log, 'Briggs log' or 'Briggsian log', too.
It is said that Henry Briggs did much work regarding logarithms to base 10. Henry was a British astronomer and mathematician specialized in geometry. So he must have been very much interested in logarithms since astronomers usually handle numbers with excessively large magnitudes. Historically though, a British mathematician, John Napier is said to have first found logarithms in early 17th century.

Besides, many scientists often use a special base, called Euler's number denoted by *e*.

Such a logarithm is called a *natural logarithm*, briefly called a natural log, too, of course.

In such a case, we usually use as the sign, **Ln** or **ln** instead of **log**.

Using a natural log, we do not usually specify the base e as in the case of a common log.   For instance, **Ln 7 = log$_e$7.**   What then, is the Euler's number e?

It is an irrational number, and is conceptually $(1 + \boldsymbol{0})^\infty$ where $\boldsymbol{0}$ is called an infinitesimal, which is as good as zero but not zero itself, and ∞ is called infinity.

Actually, $e = \sum\limits_{n=1}^{\infty} \dfrac{1}{n!} = 1 + \dfrac{1}{2} + \dfrac{1}{6} + \dfrac{1}{24} + ...$, which converges to $2.718181828459045...$,

which is an irrational number. Where do we use such a number, though?

We'll get to see a lot of es working in calculus.

9. Identities on Logarithms

In this section, we are going to take a look at some math tools that can make life easier. The tools are for algebra, which is on logarithms. So using the tools, we can handle better and work faster with expressions when doing log algebra.

The tools are called identities, which are on logarithms, and thus, called *log identities*. And we are going to see how the tools can get made, and how they work.

Talking about logs, we are talking about exponents. And in fact, a log is an exponent. We can expect therefore, that the rules or principles on exponential identities apply to log identities, too. So let's begin with some exponential identities as follows.

Suppose that both a and $b > 0$. Then, we get:

0. $a^m a^n = a^{m+n}$

1. $a^m \div a^n = a^m / a^n = \dfrac{a^m}{a^n} = a^{m-n} = 1/a^{n-m} = \dfrac{1}{a^{n-m}}$

2. $(a^m)^n = a^{mn}$

3. $(ab)^n = a^n b^n$

4. $(b/a)^n = \left(\dfrac{b}{a}\right)^n = \dfrac{b^n}{a^n} = b^n / a^n$

Let's now consider a log identity for each of the identities above.

• Beginning with the exponential identity **2** above, we have: $(a^m)^n = a^{mn}$.

What then, can we see from the log's perspective?

Assuming first, A and $b > 0$, but $b \neq 1$, we can get: $x = \log_b A$.

Then, we can get: $A = b^x$ by the definition for logs.

That's because the definition is: $A = b^x \Leftrightarrow x = \log_b A$, which means:

We get: $A = b^x \Rightarrow x = \log_b A$, and also, can get: $x = \log_b A \Rightarrow A = b^x$, too.

Next, assuming y is real, and applying the exponential identity above, we can get:

$A^y = (b^x)^y = b^{xy} = b^{yx} \Rightarrow A^y = b^{yx}$.

Next, by the definition for logs again, we can get: $yx = \log_b A^y$.

And we know: $x = \log_b A$. So we can get: $yx = y \log_b A$, too.   What then, can we get?

We have: $yx = \log_b A^y$.   So we get: $yx = y \log_b A = \log_b A^y$.

That is to say that we can get: $y \log_b A = \log_b A^y$, which is usually taken as a log identity.

The identity is saying that we get: $y \log_b A \Rightarrow \log_b A^y$, and also, get: $\log_b A^y \Rightarrow y \log_b A$, too.

For instance: we can have: $\log_3 8^2 = 2 \log_3 8$.

And we know: $8 = 2^3$. So we can get: $2 \log_3 8 = 2 \log_3 2^3 = 2 \cdot 3 \log_3 2 = 6 \log_3 2$.

In other words, we can have: $\log_3 8^2 = \log_3 (2^3)^2 = \log_3 2^6 = 6 \log_3 2$.

And for another instance, setting $A = b$ in the identity above, we get: $y \log_b b = \log_b b^y$.

So we get: $y = \log_b b^y$, because $\log_b b = 1$. How come though, we can have: $\log_b b = 1$?

We know: $b = b^1$. So we get: $1 = \log_b b$ by the definition for logs.

And thus, we now have a log identity where: $n \log_b M = \log_b M^n$.

So if the antilog is a power, we can put in front of the log the exponent of the power.

- Next, moving on to the exponential identity **0**, which is: $a^m a^n = a^{m+n}$, we can expect that assuming M, N, and $b > 0$, but $b \neq 1$, we can get: $\log_b M + \log_b N = \log_b MN$.

That's because we add together exponents when taking a product of powers with the same base.   For instance, we can get: $2^3 2^4 = 2^{2+3} = 2^7$.

Let's see now though, how we can get: $\log_b M + \log_b N = \log_b MN$.

To begin with, assuming $A = \log_b M$, and $B = \log_b N$, and using the definition for logs, we can get:

$A = \log_b M \Rightarrow M = b^A$, and $B = \log_b N \Rightarrow N = b^B$.

So we can get: $MN = b^A b^B = b^{(A+B)}$, which is: $MN = b^{(A+B)}$, which is a power.

Thus next, using the definition for logs again, we get: $A + B = \log_b MN$.

And we know: $A = \log_b M$, and $B = \log_b N$. So we get: $A + B = \log_b M + \log_b N$, too.

So we get: $\log_b M + \log_b N = \log_b MN$, which is in fact, another log identity often used.

Thus, taking <u>the sum of logs with the same base</u>, we can take the <u>product of the antilogs</u>, and then, take the log of the product to the same base. And vice versa.

So we can put a particular log into a sum of logs, where the product of the antilogs is the antilog in the particular log.   For instance:

$\log_3 9 + \log_3 27 = \log_3 9{\cdot}27 = \log_3 243$.   In fact, we have: $\log_3 9 = \log_3 3^2 = 2\log_3 3 = 2$ since $\log_b b = 1$, $3 = \log_3 27$ since $3^3 = 27$, and $\log_3 243 = \log_3 3^5 = 5\log_3 3 = 5$.

$\log_3 19683 = \log_3 9{\cdot}27{\cdot}81 = \log_3 9 + \log_3 27 + \log_3 81$.   And in fact, we have: $\log_3 9 = 2$, $\log_3 27 = \log_3 3^3 = 3$, $\log_3 81 = \log_3 3^4 = 4$, and $\log_3 19683 = 9$ since $19683 = 3^9$.

Now, getting back to the exponential identity **0** above, we can say that taking a product of powers with the same base, we take the sum of the exponents keeping base intact.

For instance, $3^2 3^3 = 3^{2+3} = 3^5$, and $3^2 3^3 3^4 = 3^{2+3+4} = 3^9$.

Note that $\log_b PQ = \log_b (PQ)$, and $\log_c PQR = \log_c (PQR)$, but that $\log_k x^2 + 2y \neq \log_k (x^2 + 2y)$.

- Next, moving on to the exponential identity **1**, we have: $a^m/a^n = \dfrac{a^m}{a^n} = a^{m-n}$.

What then, can we see from the log's perspective?

Assuming M, N, and $b > 0$, but $b \neq 1$, we can expect that: $\log_b M - \log_b N = \log_b \frac{M}{N}$.

It seems to be the case, since we take the difference between exponents doing a division with powers with the same base. For instance, we can get: $2^4/2^3 = 2^{4-3} = 2^1$.

So let's see now, if it really is the case where we can get: $\log_b M - \log_b N = \log_b \frac{M}{N}$.

To begin with, assuming $A = \log_b M$, and $B = \log_b N$, and using the definition for logs, we can get:

$A = \log_b M \Rightarrow M = b^A$, and $B = \log_b N \Rightarrow N = b^B$.

So we can get: $\dfrac{M}{N} = \dfrac{b^A}{b^B} = b^{(A-B)}$. Thus, we get: $\dfrac{M}{N} = b^{(A-B)}$, which is a power.

Thus next, using the definition for logs again, we get: $A - B = \log_b \frac{M}{N}$.

And we know: $A = \log_b M$, and $B = \log_b N$. So we get: $A - B = \log_b M - \log_b N$, too.

Thus, we get: $\log_b M - \log_b N = \log_b \frac{M}{N}$, which is in fact, another log identity often used.

So <u>subtracting a log from another to the same base</u>, we can take a <u>fraction made of the antilogs</u>, and then, take the log of the fraction to the same base. And vice versa.

Taking thus, a log of a fraction to a base, we can subtract the log of the denominator from the log of the numerator to the same base.

So for instance:

$$\log_2 8 - \log_2 16 = \log_2 \tfrac{8}{16} = \log_2 \tfrac{1}{2}. \quad \log_{0.2} \tfrac{0.04}{125} = \log_{0.2} 0.04 - \log_{0.2} 125 = 2 - (-3) = 5.$$

$$125 = 0.2^{-3}, \text{ and } \tfrac{0.04}{125} = 0.000032 = 0.2^5.$$

• Suppose next, a, b, C, and $D > 0$, but a and $b \neq 1$, and we have: $\log_a C = \log_b D$.

Then, we get: $\log_a C = \log_b D = \log_{ab} CD$.

Let's now, confirm that it really is the case.

First, setting: $\log_a C = \log_b D = k$, we get: $C = a^k$ and $D = b^k$ by the definition for logs.

Next, looking at the exponential identity **3**, we have: $(ab)^n = a^n b^n$.

62

Thus, we get: $CD = a^k b^k = (ab)^k$. So we get: $k = \log_{ab} CD$ by the definition for logs.

And we know: $\log_a C = \log_b D = k$. So we can get: $\log_a C = \log_b D = \log_{ab} CD$.

And of course, we can take it as a log identity.

In exponential algebra, taking a product of antilogs with different bases and the same exponents, we take the product of the bases, and then, use the product as the new base keeping the exponent intact. For instance, $2^3 3^3 = (2 \cdot 3)^3 = 6^3$.

So in log algebra, we can have: $3 = \log_2 2^3 = \log_3 3^3 \Rightarrow \log_2 8 = \log_3 27 = \log_6 8 \cdot 27 = 3$.

• Suppose next, a, b, C, and $D > 0$, but a and $b \neq 1$, and we have: $\log_a C = \log_b D$.

Then, we can get: $\log_a C = \log_b D = \log_{\frac{a}{b}} \frac{C}{D} = \log_{\frac{b}{a}} \frac{D}{C}$.   How come?

In exponential algebra, we have: $(\frac{b}{a})^n = \frac{b^n}{a^n}$ for $a \neq 0$, and $(\frac{a}{b})^n = \frac{a^n}{b^n}$ for $b \neq 0$.

So let's see now, if the same is true for the case of logs, too.

First, setting: $\log_a C = \log_b D = k$, we get: $C = a^k$ and $D = b^k$ by the definition for logs.

So we get: $\frac{C}{D} = \frac{a^k}{b^k} = (\frac{a}{b})^k$, and also, $\frac{D}{C} = \frac{b^k}{a^k} = (\frac{b}{a})^k$.

Thus, we get: $k = \log_{\frac{a}{b}} \frac{C}{D}$ by the definition for logs, and $k = \log_{\frac{b}{a}} \frac{D}{C}$ by the definition, too.

And we have: $\log_a C = \log_b D = k$.   So we can get: $\log_a C = \log_b D = \log_{\frac{a}{b}} \frac{C}{D} = \log_{\frac{b}{a}} \frac{D}{C}$.

In exponential algebra, doing a division with two antilogs, we divide one base by the other if the exponents are the same, and then, use the quotient as the new base keeping the exponent intact, of course. For instance: $\frac{4^3}{2^3} = (\frac{4}{2})^3 = 2^3$.

In log algebra, we can have: $3 = \log_4 4^3 = \log_2 2^3$.   And we know: $4^3 = 64$, and $2^3 = 8$.

So we get: $\log_4 64 = \log_2 8 = \log_{\frac{4}{2}} \frac{64}{8} = \log_2 8 = \log_{\frac{2}{4}} \frac{8}{64} = \log_{\frac{1}{2}} \frac{1}{8} = \log_{\frac{1}{2}} (\frac{1}{2})^3 = 3$.

So we can see that the same rules or principles on exponents do apply to logarithms, too, which sounds quite natural, since in nature, logs are no different than exponents.

Now, let's take a look at some more log identities, which are more intrinsic to the idea of logarithms.

So let's begin with assuming that A, b, and $c > 0$, but A, b, and $c \neq 1$.   Then, we get:

5. $\log_b b = 1$, and $\log_b 1 = 0$.

6. $\log_b A = \dfrac{\log_c A}{\log_c b}$.

7. $\log_b A = \dfrac{1}{\log_A b}$.

Now, we want to see how the identities above can hold, don't we?

Then again, we may want to first recall the short definition for logs as follows:

Assuming x and $y > 0$, but $x \neq 1$, we get: $\log_x y = z \Leftrightarrow y = x^z$.

5. $\log_b b = 1$, and $\log_b 1 = 0$.

To begin with, we get: $\log_b b = 1 \Leftrightarrow b^1 = b$ by the definition for logs.

And next, we get: $\log_b 1 = 0 \Leftrightarrow b^0 = 1$ by the definition, too.

6. $\log_b A = \dfrac{\log_c A}{\log_c B}$.

Doing log algebra, we often need to put a log in a fractional form.

In addition, we frequently need to modify logs so that they have the same base.

And we can quickly do so by means of the log identity above.

To begin with, suppose that $m = \log_b A$.

Then, we can see that $b^m = A$ by the definition for logs.

Next, assuming that $c > 0$ but $\neq 1$, let's take a log sub c of each side in the equality above.

That is, we take $\log_c$ of each of the two sides of the equality $b^m = A$.
How can we do that though?

As far as we do the same to both sides of an equality, the equality remains the same. That is, it still holds.   For instance, $A = B \Rightarrow A + 1 = B + 1, \frac{A}{2} = \frac{B}{2}, 3A = 3B$, etc.

By the same token, if we take the same log of each side of an equation or equality, the equal sign remains valid.   What then, do we mean by taking the same log of each side?

Taking a log of each side to the same base, we take the same log of each side.
Note that though, we don't want to use a negative number or 1 as a base of any log.

Suppose for instance, $b > 0$ but $\neq 1$, and $U = V$.

Then, taking a $\log_b$ of each side of the equality $U = V$, we get: $\log_b U = \log_b V$.

We know for instance, $100 = 100$.   So taking $\log_{10}$ of each of the two sides, we get:

$\log_{10} 100 = \log_{10} 100 \Rightarrow \log_{10} 10^2 = \log_{10} 10^2 \Rightarrow 2 \log_{10} 10 = 2 \log_{10} 10 \Rightarrow 2 = 2$.

So we can say that '$\log_b$' is a *log operator*. And of course, the base b can be any number that can be a base of a log. So the base b has to be > 0, but $\neq 1$.

If for instance, we take a log of N to a base b, we can say that $\log_b$ is applied to N, so $\log_b$ is the operator, and N is the operand.   What then, do we get?

Taking a log of N to a base b, we get: $\log_b N$, which is an exponent to which the base b is raised to equal the number N. For instance, taking **log** of 9 to base 3, we apply $\log_3$ to 9, and we get $\log_3 9$, which is 2, because $9 = 3^2$.   And we can put it the way below, too:

Taking **log sub 3** of 9, we apply $\log_3$ to 9, and we get $\log_3 9$, which is 2.

Thus, $\log_3 9$ can be read as **log sub 3 of 9** as well as **log of 9 to base 3**, or **log of 9 to 3**.

Now, let's get back to the point where we left off.
We are taking $\log_c$ of each side of the equality where $b^m = A$.

Then, we get: $\log_c b^m = \log_c A$. And next, we know: $\log_c b^m = m \log_c b$.

So we get: $m \log_c b = \log_c A$, since $b^m = A$.   Thus, we get: $m = \dfrac{\log_c A}{\log_c B}$.

And we have: $m = \log_b A$, too.   Therefore, we can see that $\log_b A = \dfrac{\log_c A}{\log_c b}$.

For instance, in exponential algebra, we can have: $27 = \frac{81}{3} = \frac{9^2}{9^{\frac{1}{2}}} = 9^{2-\frac{1}{2}} = 9^{\frac{3}{2}}$.

Thus, in log algebra, we can get: $\log_9 27 = \log_9 9^{\frac{3}{2}} = \frac{3}{2}$, and also, $\log_9 27 = \frac{\log_3 27}{\log_3 9} = \frac{3}{2}$.

In short:

Suppose first, $m = \log_b A$. Then, we get: $b^m = A$ by the definition for logs.
Suppose next, that $c > 0$ and $\neq 1$, and that we take $\log_c$ of each of the two sides of the equality where $b^m = A$.   Then, we get: $\log_c b^m = \log_c A$.

We know: $\log_c b^m = m \log_c b$. So we get: $m \log_c b = \log_c A$. Then, we get: $m = \dfrac{\log_c A}{\log_c b}$.

And we have: $m = \log_b A$, too.   Therefore, we can get: $\log_b A = \dfrac{\log_c A}{\log_c b}$.

Note that the base c above can be any number if the number can be a base in a log. And we know that a <u>base in a log</u> has to be <u>positive but unequal to 1</u>.

Then, can we use the base b in the $\log_b A$ as the base c in the $\log_c b$, too?

Yes, we can. In $\dfrac{\log_c A}{\log_c b}$, replacing c with b, we get: $\log_b A = \frac{\log_b A}{\log_b b} = \frac{\log_b A}{1} = \log_b A$.

Well, the replacement with b doesn't give us much of significance, since it doesn't make any difference. Nevertheless, it still works. That is, the equal sign remains valid.

Then, can we use the antilog A in the $\log_b A$ as the base c in the $\log_c b$, too?

Yes, we can **unless** A is equal to 1.   What if A is not positive?

A is already an antilog in a log, so it is positive. We'll see how it can be the case in the next log identity, which is the identity numbered **7**, and will be explained shortly.

Changing the base in a log keeping intact the value of the log, we usually put the log in such a fractional form as described above.

Suppose for instance, A, b, c, and $d > 0$, but b, c, and $d \neq 1$. Then, we can get:

$\log_b A = \frac{\log_c A}{\log_c b} = \frac{\log_d A}{\log_d b} = \frac{\log_3 A}{\log_3 b} = \frac{\log_{0.3} A}{\log_{0.3} b} = \frac{\log_\pi A}{\log_\pi b} = \ldots$ where π is the circular ratio, which is

3.141592…, which is an irrational number.

By the way, in a natural log, where we use as the log sign, **ln** (or **Ln**) instead of **log**, we use as the base, the Euler's number, denoted by e. The number e is an irrational number, and is approximately 2.72. To be more accurate, it is approximately 2.718181828459045.

So we can set: $\log_b A = \dfrac{\log_e A}{\log_e b}$, too, which is normally put in $\dfrac{\ln A}{\ln b}$.

Note:

Suppose A, b, c, S, and $t > 0$, but b, c, and $t \neq 1$.

Then, $\log A$ is a common log of A, and is the same as $\log_{10} A$. So we get: $\log 10 = 1$.

And we have a log identity where: $\log_b A = \dfrac{\log_c A}{\log_c b}$.

So we get: $\log_b A = \dfrac{\log_{10} A}{\log_{10} b} = \dfrac{\log A}{\log b}$, and thus, we get: $\log_b A = \dfrac{\log A}{\log b}$.

By the same token, we get: $\dfrac{\log S}{\log t} = \log_t S$.

7. $\log_b A = \dfrac{1}{\log_A b}$ where A and b both are > 0 but $\neq 1$.

Basically, this log identity is no different from the one explained above.

Suppose now, that $m = \log_b A$. Then, we get: $b^m = A$ by the definition for logs.

Next, we know that $A > 0$ but $\neq 1$.

So taking $\log_A$ of each side of $b^m = A$, we get: $\log_A b^m = \log_A A = 1 \Rightarrow \log_A b^m = 1$.

And we have $\log_A b^m = m \log_A b$, too.

So we can see that $\log_A b^m = 1 \Rightarrow m \log_A b = 1 \Rightarrow m = \dfrac{1}{\log_A b}$.

And we have: $m = \log_b A$. Therefore, we can see that $\log_b A = \dfrac{1}{\log_A b}$.

In short:

$$\log_b A = \frac{\log_A A}{\log_A b} = \frac{1}{\log_A b}.$$

Note:

Suppose $A > 0$.

Then, $\ln A$ is a natual log of A, and is the same as $\log_e A$ where e is the Euler's number. We can have $\ln e$, which is 1, because $\ln e = \log_e e = 1$.

However, we do not have such a log as $\ln_b A$, which does not make sense.

So we have: $\log_b A = \dfrac{\log_e A}{\log_e b} = \dfrac{\ln A}{\ln b}$, which is however, not $\ln_b A$, which doesn't exist.

Now, putting threads together, we have:

$y \log_b A = \log_b A^y.$

$\log_b M + \log_b N = \log_b MN.$ $\qquad$ $\log_b M - \log_b N = \log_b \frac{M}{N}.$

$\log_a C = \log_b D = \log_{ab} CD.$ $\qquad$ $\log_a C = \log_b D = \log_{\frac{a}{b}} \frac{C}{D} = \log_{\frac{b}{a}} \frac{D}{C}.$

$\log_b b = 1$, and $\log_b 1 = 0.$

$\log_b A = \dfrac{\log_c A}{\log_c b}.$ $\qquad$ $\log_b A = \dfrac{1}{\log_A b}.$

Examples 1 on Log Identities

We have some log identities as follows:

$y \log_b A = \log_b A^y$.   So for instance, $2 \log_3 9 = \log_3 9^2$.

$\log_b M + \log_b N = \log_b MN$.   So for instance, $\log_3 9 + \log_3 27 = \log_3 9 \cdot 27 = \log_3 243$.

$\log_b M - \log_b N = \log_b \frac{M}{N}$.   So for instance, $\log_3 27 - \log_3 9 = \log_3 27/9 = \log_3 3$.

$\log_a C = \log_b D = \log_{ab} CD$.   So for instance, $\log_3 9 = \log_2 4 = \log_6 36$.

$\log_a C = \log_b D = \log_{\frac{a}{b}} \frac{C}{D} = \log_{\frac{b}{a}} \frac{D}{C}$.   So for instance, $\log_4 16 = \log_2 4 = \log_2 4 = \log_{\frac{1}{2}} \frac{1}{4}$.

$\log_b b = 1$, and $\log_b 1 = 0$.   So for instance, $\log_{10} 10 = \log_2 2 = 1$, and $\log_7 1 = \log_3 1 = 0$.

$\log_b A = \dfrac{\log_c A}{\log_c b}$.   So for instnace, $\log_4 16 = \dfrac{\log_2 16}{\log_2 4} = \dfrac{\log_3 16}{\log_3 4} = \dfrac{\log_4 16}{\log_4 4} = \dfrac{\log_5 16}{\log_5 4} = \ldots$

$\log_b A = \dfrac{1}{\log_A b}$.   So for instance, $\log_4 16 = \dfrac{1}{\log_{16} 4}$.

Now, do the examples as below:

0. Assuming $\log_{10} 2 = 0.301$, and $\log_{10} 3 = 0.4771$, find the value of $\log_2 54$.

1. Assuming $\log_2 12 = 3.587$, find the value of $\log_2 54$.

2. Assuming $\log_2 12 = x$, express $\log_2 54$ in terms of x.

3. Assuming $x = \log_b \frac{16}{9}$, and $y = \log_b \frac{27}{8}$, express $\log_b 24$ in terms of x and y.

4. Assuming $a^x b^y = b$ where a and $b > 0$ but $a \neq 1$, express $\log_a a^y b^x$ in terms of x and y.

5. Assuming $a^3 b^5 = b$ where a and $b > 0$ but $a \neq 1$, find the value of $\log_a a^5 b^3$.

6. Assuming $\frac{a^x}{b^y} = a$ where a and $b > 0$ but $b \neq 1$, express $\log_b \frac{a^y}{b^x}$ in terms of x and y.

7. Assuming $\frac{a^3}{b^5} = a$ where a and $b > 0$ but $b \neq 1$, find the value of $\log_b \frac{a^5}{b^3}$.

8. Find the value of $\log_2 2\sqrt{6} - \frac{1}{2}\log_2 \frac{1}{7} - \frac{5}{2}\log_2 \sqrt[5]{21}$.

Suggestions or Solutions
To the Examples 1 on Log Identities

0.   Assuming $\log_{10} 2 = 0.301$, and $\log_{10} 3 = 0.4771$, find the value of $\log_2 54$.

First, we have a log identity where $\log_b A = \dfrac{\log_c A}{\log_c b}$. So we can set: $\log_2 54 = \dfrac{\log_{10} 54}{\log_{10} 2}$.

Next, we have another log identity where $\log_b M + \log_b N = \log_b MN$.

So we can get: $\log_{10} 54 = \log_{10} 2 \cdot 27 = \log_{10} 2 + \log_{10} 27$, where $\log_{10} 2 = 0.301$.

And next, we have another log identity where $y \log_b A = \log_b A^y$.

So we can get: $\log_{10} 27 = \log_2 3^3 = 3 \log_2 3$, where $\log_{10} 3 = 0.4771$.

Thus, we get: $\log_{10} 2 + \log_{10} 27 = \log_{10} 2 + 3 \log_2 3 = 0.301 + 3 \cdot 0.4771 = 1.7323$.

In sum, we get:

$\log_{10} 54 = \log_{10} 2 \cdot 27 = \log_{10} 2 + \log_{10} 27 = \log_{10} 2 + \log_2 3^3 = \log_{10} 2 + 3 \log_2 3$

$= 0.301 + 3 \cdot 0.4771 = 1.7323$.

So we get: $\log_2 54 = \dfrac{\log_{10} 54}{\log_{10} 2} = \dfrac{1.7323}{0.301} = 5.7551$.

1.   Assuming $\log_2 12 = 3.587$, find the value of $\log_2 54$.

$\log_2 12 = \log_2 3 \cdot 4 = \log_2 3 + \log_2 4 = \log_2 3 + \log_2 2^2 = \log_2 3 + 2 \log_2 2 = \log_2 3 + 2 \cdot 1$

$= 3.587 \Rightarrow \log_2 3 = 3.587 - 2 = 1.587 \Rightarrow \log_2 3 = 1.587$.

$\log_2 54 = \log_2 2 \cdot 27 = \log_2 2 + \log_2 27 = 1 + \log_2 3^3 = 1 + 3 \log_2 3 = 1 + 3 \cdot 1.587 = 5.761$.

2. Assuming $\log_2 12 = x$, express $\log_2 54$ in terms of x.

$\log_2 12 = \log_2 3 \cdot 4 = \log_2 3 + \log_2 4 = \log_2 3 + 2 = x \Rightarrow \log_2 3 = x - 2$.

$\log_2 54 = \log_2 2 \cdot 27 = \log_2 2 + \log_2 27 = 1 + 3 \log_2 3 = 1 + 3x - 6 = 3x - 5$.

3. Assuming $x = \log_b \frac{16}{9}$, and $y = \log_b \frac{27}{8}$, express $\log_b 24$ in terms of x and y.

First, $\log_b 24 = \log_b 3 \cdot 8 = \log_b 3 + \log_b 8 = \log_b 3 + \log_b 2^3 = \log_b 3 + 3 \log_b 2$.

Next, we have a log identity where $\log_b M - \log_b N = \log_b \frac{M}{N}$.

So we can get: $x = \log_b \frac{16}{9} = \log_b 16 - \log_b 9 = \log_b 2^4 - \log_b 3^2 = 4 \log_b 2 - 2 \log_b 3$, and

$y = \log_b \frac{27}{8} = \log_b 27 - \log_b 8 = \log_b 3^3 - \log_b 2^3 = 3 \log_b 3 - 3 \log_b 2$.

So next, setting: $u = \log_b 2$, and $v = \log_b 3$, we have:

$\log_b 24 = v + 3u$, $x = 4u - 2v$, and $y = 3v - 3u$.

How then, can we put $\log_b 24$ in terms of x and y?

We can take the last two equations as a system for u and v.
So solving it, and then, putting the solution into the first equation, we can put $\log_b 24$ in terms of x and y.

So to begin with, solving the system where $x = 4u - 2v$ and $y = 3v - 3u$, we can put them this way first: $3x = 12u - 6v$ and $2y = 6v - 6u$.

Thus next, we can get: $3x + 2y = 12u - 6v + 6v - 6u = 6u \Rightarrow u = \frac{1}{6}(3x + 2y)$.

So next, putting u into either of the two equations in the system, we can get v.

Using $x = 4u - 2v$, we get: $2v = 4u - x = \frac{4}{6}(3x + 2y) - x = 2x + \frac{4}{3}y - x = x + \frac{4}{3}y$.

So we get: $v = \frac{1}{2}x + \frac{2}{3}y$.

Thus, we get: $\log_b 24 = v + 3u = \frac{1}{2}x + \frac{2}{3}y + \frac{1}{2}(3x + 2y) = 2x + \frac{5}{3}y$.

4.   Assuming $a^x b^y = b$ where a and $b > 0$ but $a \neq 1$, express $\log_a a^y b^x$ in terms of x and y.

To begin with, we can get: $\log_a a^y b^x = \log_a a^y + \log_a b^x = y \log_a a + x \log_a b = y + x \log_a b$.

So we get: $\log_a a^y b^x = y + x \log_a b$.   What then, do we want to get?

We want to put $\log_a b$ in terms of x and y.   How though?

We have: $a^x b^y = b$ where a and $b > 0$ but $a \neq 1$,

So taking first, the log sub a of both sides in $a^x b^y = b$, we get: $\log_a a^x b^y = \log_a b$.

Next, we can get: $\log_a a^x b^y = \log_a a^x + \log_a b^y = x \log_a a + y \log_a b = x + y \log_a b$.

So we get: $x + y \log_a b = \log_a b \Rightarrow x + y \log_a b - \log_a b = 0 \Rightarrow x + (\log_a b)(y - 1) = 0$.

Thus, we get: $\log_a b = \dfrac{x}{1 - y}$.

So we get: $\log_a a^y b^x = y + x \log_a b = y + x \cdot \dfrac{x}{1 - y} = y + \dfrac{x^2}{1 - y} = \dfrac{y - y^2 + x^2}{1 - y}$.

5. Assuming $a^3 b^5 = b$ where a and $b > 0$ but $a \neq 1$, find the value of $\log_a a^5 b^3$.

To begin with, we can get: $\log_a a^5 b^3 = \log_a a^5 + \log_a b^3 = 5 \log_a a + 3 \log_a b = 5 + 3 \log_a b$.

So we get: $\log_a a^5 b^3 = 5 + 3 \log_a b$. And thus, we want to get the value of $\log_a b$.

First, we have: $a^3 b^5 = b$ where a and $b > 0$ but $a \neq 1$.

So taking next, the log sub a of both sides in $a^3 b^5 = b$, we get: $\log_a a^3 b^5 = \log_a b$.

Next, we can get: $\log_a a^3 b^5 = \log_a a^3 + \log_a b^5 = 3 \log_a a + 5 \log_a b = 3 + 5 \log_a b$.

So we get: $3 + 5 \log_a b = \log_a b \Rightarrow 3 + 5 \log_a b - \log_a b = 0 \Rightarrow 3 + 4 \log_a b = 0$.

Thus, we get: $\log_a b = -3/4$. So we get: $\log_a a^5 b^3 = 5 + 3 \log_a b = 5 - 3 \cdot (3/4) = 11/4$.

6. Assuming $\frac{a^x}{b^y} = a$ where a and $b > 0$ but $b \neq 1$, express $\log_b \frac{a^y}{b^x}$ in terms of x and y.

First, we can get: $\log_b \frac{a^y}{b^x} = \log_b a^y - \log_b b^x = y \log_b a - x \log_b b = y \log_b a - x$.

What then, do we want to get?

We want to put $\log_b a$ in terms of x and y. How?

We have: $\frac{a^x}{b^y} = a$ where a and $b > 0$ but $b \neq 1$,

So taking first, the log sub b of both sides in $\frac{a^x}{b^y} = a$, we get: $\log_b \frac{a^x}{b^y} = \log_b a$.

Next, we can get: $\log_b \frac{a^x}{b^y} = \log_b a^x - \log_b b^y = x \log_b a - y \log_b b = x \log_b a - y$.

(Note that $\log_b (a - y) \neq \log_b a - y$.)

So we get: $x \log_b a - y = \log_b a \Rightarrow x \log_b a - y - \log_b a = 0 \Rightarrow (\log_b a)(x - 1) - y = 0$.

Thus, we get: $\log_b a = \frac{y}{x-1}$.

So we get: $\log_b \frac{a^y}{b^x} = y \log_b a - x = y \cdot \frac{y}{x-1} - x = \frac{y^2}{x-1} - x = \frac{y^2 - x^2 + x}{x-1}$.

7.   Assuming $\frac{a^3}{b^5} = a$ where a and $b > 0$ but $b \neq 1$, find the value of $\log_b \frac{a^5}{b^3}$.

First, we can get: $\log_b \frac{a^5}{b^3} = \log_b a^5 - \log_b b^3 = 5 \log_b a - 3 \log_b b = 5 \log_b a - 3$.

So next, we want to get the value of $\log_b a$.

We have: $\frac{a^3}{b^5} = a$ where a and $b > 0$ but $b \neq 1$,

So taking first, the log sub b of both sides in $\frac{a^3}{b^5} = a$, we get: $\log_b \frac{a^3}{b^5} = \log_b a$.

Next, we can get: $\log_b \frac{a^3}{b^5} = \log_b a^3 - \log_b b^5 = 3 \log_b a - 5 \log_b b = 3 \log_b a - 5$.

So we get: $3 \log_b a - 5 = \log_b a \Rightarrow 3 \log_b a - 5 - \log_b a = 0 \Rightarrow 2 \log_b a - 5 = 0$.

Thus, we get: $\log_b a = 5/2$.

So we get: $\log_b \frac{a^5}{b^3} = 5 \log_b a - 3 = 5 \cdot \frac{5}{2} - 3 = \frac{25}{2} - 3 = \frac{19}{2}$.

8.   Find the value of $\log_2 2\sqrt{6} - \frac{1}{2}\log_2 \frac{1}{7} - \frac{5}{2}\log_2 \sqrt[5]{21}$.

$$\log_2 2\sqrt{6} = \log_2 2 + \log_2 \sqrt{6} = 1 + \log_2 \sqrt{6}$$

$$\frac{1}{2}\log_2 \frac{1}{7} = \log_2 \left(\frac{1}{7}\right)^{\frac{1}{2}} = \log_2 \sqrt{\frac{1}{7}} = \log_2 \frac{1}{\sqrt{7}}$$

$$\frac{5}{2}\log_2 \sqrt[5]{21} = \frac{5}{2}\log_2 21^{\frac{1}{5}} = \frac{5}{2} \cdot \frac{1}{5}\log_2 21 = \frac{1}{2}\log_2 21 = \log_2 \sqrt{21}$$

So we get:

$$\log_2 2\sqrt{6} - \frac{1}{2}\log_2 \frac{1}{7} - \frac{5}{2}\log_2 \sqrt[5]{21} = 1 + \log_2 \sqrt{6} - \log_2 \frac{1}{\sqrt{7}} - \log_2 \sqrt{21}$$

$$= 1 + \log_2 \sqrt{6} + \log_2 \sqrt{7} - \log_2 \sqrt{21} = 1 + \log_2 \frac{\sqrt{6}\sqrt{7}}{\sqrt{21}} = 1 + \log_2 \frac{\sqrt{2}\sqrt{3}\sqrt{7}}{\sqrt{21}} = 1 + \log_2 \sqrt{2} = \frac{3}{2}.$$

Examples 2 on Log Identities

To begin with, we have some log identities as follows:

$y \log_b A = \log_b A^y$.   So for instance, $2 \log_3 9 = \log_3 9^2$.

$\log_b M + \log_b N = \log_b MN$.   So for instance, $\log_3 9 + \log_3 27 = \log_3 9 \cdot 27 = \log_3 243$.

$\log_b M - \log_b N = \log_b \frac{M}{N}$.   So for instance, $\log_3 27 - \log_3 9 = \log_3 27/9 = \log_3 3$.

$\log_a C = \log_b D = \log_{ab} CD$.   So for instance, $\log_3 9 = \log_2 4 = \log_6 36$.

$\log_a C = \log_b D = \log_{\frac{a}{b}} \frac{C}{D} = \log_{\frac{b}{a}} \frac{D}{C}$.   So for instance, $\log_4 16 = \log_2 4 = \log_2 4 = \log_{\frac{1}{2}} \frac{1}{4}$.

$\log_b b = 1$, and $\log_b 1 = 0$.   So for instance, $\log_{10} 10 = \log_2 2 = 1$, and $\log_7 1 = \log_3 1 = 0$.

$\log_b A = \dfrac{\log_c A}{\log_c b}$.   So for instnace, $\log_4 16 = \dfrac{\log_2 16}{\log_2 4} = \dfrac{\log_3 16}{\log_3 4} = \dfrac{\log_4 16}{\log_4 4} = \dfrac{\log_5 16}{\log_5 4} = \ldots$

$\log_b A = \dfrac{1}{\log_A b}$.   So for instance, $\log_4 16 = \dfrac{1}{\log_{16} 4}$.

Now, do the examples as below:

0. Assuming $a > b > 0$, simplify the expression below:

$$\sqrt{\log_c \sqrt{ab} + \sqrt{\log_c a \cdot \log_c b}} + \sqrt{\log_c \sqrt{ab} - \sqrt{\log_c a \cdot \log_c b}}$$

1. Assuming $x > 0$, simplify $\sqrt{\log_2 8x - \sqrt{\log_2 x^{12}}}$.

2. Assuming $x = \frac{1}{36}(9^{\frac{1}{y}} + 9^{-\frac{1}{y}})$, find the value of $y \log_9 3(\sqrt{x + \frac{1}{18}} + \sqrt{x - \frac{1}{18}})$.

3. Assuming $x > 1$, put $\sqrt{\log 10\sqrt{x} - \sqrt{\log x^2}}$ in terms of $\sqrt{\log x}$.

4. Assuming x and y are rational numbers, but $\log_3 2$ is not, and $x \log_3 2 + y \log_3 6 = 1$, find x and y.

Suggestions or Solutions
To the Problem in the Example 0

Assuming $a > b > 0$, simplify the expression below:

$$\sqrt{\log_c \sqrt{ab} + \sqrt{\log_c a \cdot \log_c b}} + \sqrt{\log_c \sqrt{ab} - \sqrt{\log_c a \cdot \log_c b}}$$

To begin with, we can get: $\log_c \sqrt{ab} = \log_c(\sqrt{a}\sqrt{b}) = \log_c \sqrt{a} + \log_c \sqrt{b}$

Next, $\sqrt{\log_c a \cdot \log_c b} = 2 \cdot \frac{1}{2}\sqrt{\log_c a \cdot \log_c b} = 2\sqrt{\frac{1}{4}(\log_c a \cdot \log_c b)} = 2\sqrt{\frac{1}{2}\log_c a \cdot \frac{1}{2}\log_c b}$

$$= 2\sqrt{\log_c \sqrt{a} \cdot \log_c \sqrt{b}}$$

So we get: $\log_c \sqrt{ab} - \sqrt{\log_c a \cdot \log_c b} = \log_c \sqrt{a} + \log_c \sqrt{b} - 2\sqrt{\log_c \sqrt{a} \cdot \log_c \sqrt{b}}$

$$= (\sqrt{\log_c \sqrt{a}} - \sqrt{\log_c \sqrt{b}})^2. \quad \text{And we know: } a > b > 0, \text{ so we get: } \sqrt{\log_c \sqrt{a}} > \sqrt{\log_c \sqrt{b}}.$$

Thus, we get: $\sqrt{\log_c \sqrt{ab} - \sqrt{\log_c a \cdot \log_c b}} = \sqrt{\log_c \sqrt{a}} - \sqrt{\log_c \sqrt{b}}$.

And by the same token, we can get: $\sqrt{\log_c \sqrt{ab} + \sqrt{\log_c a \cdot \log_c b}} = \sqrt{\log_c \sqrt{a}} + \sqrt{\log_c \sqrt{b}}$.

So assuming the expression given is P, we get: $P = 2\sqrt{\log_c \sqrt{a}} = \sqrt{4\log_c \sqrt{a}}$

$$= \sqrt{4 \cdot \frac{1}{2}\log_c a} = \sqrt{2\log_c a} = \sqrt{\log_c a^2}.$$

Suggestions or Solutions
To the Problem in the Example 1

Assuming $x > 0$, simplify $\sqrt{\log_2 8x - \sqrt{\log_2 x^{12}}}$.

First, we can get: $\log_2 x^{12} = 12\log_2 x$.

So we get: $\sqrt{\log_2 x^{12}} = \sqrt{12\log_2 x} = \sqrt{4 \cdot 3\log_2 x} = \sqrt{4}\sqrt{3\log_2 x} = 2\sqrt{3\log_2 x}$.

And next, we can get:
$\log_2 8x = \log_2 8 + \log_2 x = \log_2 2^3 + \log_2 x = 3\log_2 2 + \log_2 x = 3 + \log_2 x$.

Thus, we get: $\log_2 8x - \sqrt{\log_2 x^{12}} = 3 + \log_2 x - 2\sqrt{3\log_2 x} = 3 + \log_2 x - 2\sqrt{3}\sqrt{\log_2 x}$

$= (\sqrt{3} - \sqrt{\log_2 x})^2$.

And we know what's inside a square root sign is positive or 0.

So we get: $\sqrt{\log_2 8x - \sqrt{\log_2 x^{12}}} = \left|\sqrt{3} - \sqrt{\log_2 x}\right|$.

Suggestions or Solutions
To the Problem in the Example 2

Assuming $x = \frac{1}{36}(9^{\frac{1}{y}} + 9^{-\frac{1}{y}})$, find the value of $y \log_9 3(\sqrt{x + \frac{1}{18}} + \sqrt{x - \frac{1}{18}})$.

First, $x + \frac{1}{18} = \frac{1}{36}(9^{\frac{1}{y}} + 9^{-\frac{1}{y}}) + \frac{1}{18} = \frac{1}{36}(9^{\frac{1}{y}} + 9^{-\frac{1}{y}} + 2) = \frac{1}{36}(3^{\frac{2}{y}} + 3^{-\frac{2}{y}} + 2) = \frac{1}{36}(3^{\frac{1}{y}} + 3^{-\frac{1}{y}})^2$.

So we get: $\sqrt{x + \frac{1}{18}} = \frac{1}{6}(3^{\frac{1}{y}} + 3^{-\frac{1}{y}})$.

Next, $x - \frac{1}{18} = \frac{1}{36}(9^{\frac{1}{y}} + 9^{-\frac{1}{y}}) - \frac{1}{18} = \frac{1}{36}(9^{\frac{1}{y}} + 9^{-\frac{1}{y}} - 2) = \frac{1}{36}(3^{\frac{2}{y}} + 3^{-\frac{2}{y}} - 2) = \frac{1}{36}(3^{\frac{1}{y}} - 3^{-\frac{1}{y}})^2$.

So we get: $\sqrt{x - \frac{1}{18}} = \frac{1}{6}(3^{\frac{1}{y}} - 3^{-\frac{1}{y}})$, because $3^{\frac{1}{y}} - 3^{-\frac{1}{y}} > 0$, since $y > 0$.

Thus, we get: $3(\sqrt{x + \frac{1}{18}} + \sqrt{x - \frac{1}{18}}) = 3 \cdot \frac{1}{3} \cdot 3^{\frac{1}{y}} = 3^{\frac{1}{y}}$.

So we get: $y \log_9 3(\sqrt{x + \frac{1}{18}} + \sqrt{x - \frac{1}{18}}) = y \log_9 3^{\frac{1}{y}} = \log_9 (3^{\frac{1}{y}})^y = \log_9 3 = \log_9 9^{1/2} = 1/2$.

Suggestions or Solutions
To the Problem in the Example 3

Assuming $x > 1$, put $\sqrt{\log 10\sqrt{x} - \sqrt{\log x^2}}$ in terms of $\sqrt{\log x}$.

To begin with, $\log 10\sqrt{x} = \log 10 + \log \sqrt{x} = 1 + \frac{1}{2}\log x$, and $\log x^2 = 2 \log x$.

Thus next, we can get:

$\log 10\sqrt{x} - \sqrt{\log x^2} = 1 + \frac{1}{2}\log x - \sqrt{2\log x} = \frac{1}{2}(2 + \log x - 2\sqrt{2\log x}) = \frac{1}{2}(\sqrt{2} - \sqrt{\log x})^2$.

Meanwhile, $\sqrt{2} - \sqrt{\log x} \geq 0 \Rightarrow \sqrt{\log x} \leq \sqrt{2} \Rightarrow \log x \leq 2 \Rightarrow x \leq 10^2$.

So we get: $\sqrt{2} - \sqrt{\log x} < 0 \Rightarrow x > 10^2$. And we have: $x > 1$, too.

Thus, we get:

If $1 \leq x \leq 100$, $\sqrt{\log 10\sqrt{x} - \sqrt{\log x^2}} = \frac{1}{\sqrt{2}}(\sqrt{2} - \sqrt{\log x})$.

If $x > 10^2$, $\sqrt{\log 10\sqrt{x} - \sqrt{\log x^2}} = \frac{1}{\sqrt{2}}(\sqrt{\log x} - \sqrt{2})$.

If not quite sure of the idea behind the processes above, follow the steps below:

To begin with, we can get:

$\log 10\sqrt{x} = \log 10 + \log \sqrt{x} = 1 + \frac{1}{2}\log x$, and $\log x^2 = 2 \log x$.

And next, we have: $x > 1$. So we can get: $\sqrt{\log x}$, since $\log x > 0$.

Thus next, we can get:

$\log 10\sqrt{x} - \sqrt{\log x^2} = 1 + \frac{1}{2}\log x - \sqrt{2\log x} = \frac{1}{2}(2 + \log x - 2\sqrt{2\log x})$.

Meanwhile, we can have: $(\sqrt{2} - \sqrt{\log x})^2 = 2 + \log x - 2\sqrt{\log x}$.

So we get: $\log 10\sqrt{x} - \sqrt{\log x^2} = \sqrt{\frac{1}{2}(\sqrt{2} - \sqrt{\log x})^2} = \frac{1}{\sqrt{2}}\sqrt{(\sqrt{2} - \sqrt{\log x})^2}$.

And we know: $\sqrt{(\sqrt{2} - \sqrt{\log x})^2} \geq 0$. So we get two cases. What are the two though?

One is: $\sqrt{(\sqrt{2} - \sqrt{\log x})^2} = \sqrt{2} - \sqrt{\log x}$ if $\sqrt{2} - \sqrt{\log x} \geq 0$.

And the other is: $\sqrt{(\sqrt{2} - \sqrt{\log x})^2} = \sqrt{\log x} - \sqrt{2}$ if $\sqrt{2} - \sqrt{\log x} < 0$.

What then, is the next?

We want to get the extent of x in each case.

To begin with, we have: $x > 1$. And next, we can have:

$\sqrt{2} - \sqrt{\log x} \geq 0 \Rightarrow \sqrt{\log x} \leq \sqrt{2} \Rightarrow \log x \leq 2 \Rightarrow x \leq 10^2$. So we get: $1 \leq x \leq 100$.

$\sqrt{2} - \sqrt{\log x} < 0 \Rightarrow \sqrt{\log x} > \sqrt{2} \Rightarrow \log x > 2 \Rightarrow x > 10^2$.

Thus, we get:

If $1 \leq x \leq 100$, $\sqrt{\log 10\sqrt{x} - \sqrt{\log x^2}} = \frac{1}{\sqrt{2}}\sqrt{(\sqrt{2} - \sqrt{\log x})^2} = \frac{1}{\sqrt{2}}(\sqrt{2} - \sqrt{\log x})$.

If $x > 10^2$, $\sqrt{\log 10\sqrt{x} - \sqrt{\log x^2}} = \frac{1}{\sqrt{2}}(\sqrt{\log x} - \sqrt{2})$.

Suggestions or Solutions
To the Problem in the Example 4

Assuming x and y are rational numbers, but $\log_3 2$ is not, and $x \log_3 2 + y \log_3 6 = 1$, find x and y.

To begin with, we have: $\log_3 6 = \log_3 2 \cdot 3 = \log_3 2 + \log_3 3 = \log_3 2 + 1$.

So next, we can get: $x \log_3 2 + y \log_3 6 = x \log_3 2 + y (\log_3 2 + 1) = (x + y)\log_3 2 + y$.

And we have: $x \log_3 2 + y \log_3 6 = 1$.

So we get: $(x + y)\log_3 2 + y = 1 \Rightarrow (x + y)\log_3 2 + y - 1 = 0$.

And we know that $\log_3 2$ is not a rational number, but that x and y are.

So we get: $x + y = 0$, and $y - 1 = 0$. Thus, we get: $x = -y$, and $y = 1$.

So we get: $x = -1$, and $y = 1$.

Examples 3 on Log Identities

0.0.   Using $\log_3 6 = 1.631$, find the value of $\log_3 24$.

0.1.   Assuming $f(x) = \log_a x$, and n is constant, put each of $f(xy)$, $f\left(\frac{x}{y}\right)$, and $f(x^n)$ in terms of $f(x)$ and $f(y)$.

1.   Assuming $a^5 b^2 = 1$, find the value of $\log_a a^4 b^5$.

2.   Assuming $f(x) = 2^{\frac{x}{2}}$, evaluate $f(x)$ at $x = 2 \log_2 3 + \log_4 6 - \log_8 2\sqrt{2}$.

3.   Show that $a^{\log b} = b^{\log a}$.

4.   Assuming $b^x = u$, $b^y = v$, and $b^z = w$, express $\log_{uv^2 w^3} u^3 v^2 w$ in terms of x, y, and z.

5.   Find the ratio $a : b : c$, assuming that a, b, and $c > 0$, but $c \neq 1$, and that:

$\log_b (2b + c) + \log_b (2b - c) = 2$

$\log_{a+b} c + \log_{a-b} c = (2 \log_{a+b} c)(\log_{a-b} c)$ where $c \neq 1$

Suggestions or Solutions
To the Problems in the Example 0

0. Using $\log_3 6 = 1.631$, find the value of $\log_3 24$.

We have: $\log_3 24 = \log_3 3 + \log_3 8 = 1 + 3 \log_3 2$.
It looks like we don't have the value of $\log_3 2$, though.

Yes, we do, and it's in $\log_3 6$.   How then, can we get it?

We have: $\log_3 6 = \log_3 3 + \log_3 2 = 1 + \log_3 2 = 1.631 \Rightarrow \log_3 2 = 0.631$.

Thus, we get: $\log_3 24 = 1 + 3 \log_3 2 = 1 + 3 \cdot 0.631 = 1 + 1.893 = 2.893$.

1. Assuming $f(x) = \log_a x$, and n is constant, put each of $f(xy)$, $f\left(\frac{x}{y}\right)$, and $f(x^n)$ in terms of $f(x)$ and $f(y)$.

$f(xy) = \log_a xy = \log_a x + \log_a y = f(x) + f(y)$.

So if $f(x) = \log_a x$, we get: $f(xy) = f(x) + f(y)$.

For instance, $f(35) = \log_a 35 = \log_a 5 + \log_a 7 = f(5) + f(7)$.

$f\left(\frac{x}{y}\right) = \log_a \frac{x}{y} = \log_a x - \log_a y = f(x) - f(y)$.

So if $f(x) = \log_a x$, we get: $f\left(\frac{x}{y}\right) = f(x) - f(y)$.

For instance, $f\left(\frac{2}{5}\right) = \log_a \frac{2}{5} = \log_a 2 - \log_a 5 = f(2) - f(5)$.

$f(x^n) = \log_a x^n = n \log_a x = nf(x)$.

So if $f(x) = \log_a x$, we get: $f(x^n) = nf(x)$.

For instance, $f(2^3) = \log_a 2^3 = 3 \log_a 2 = 3f(2)$.

Suggestions or Solutions
To the Problem in the Example 1

Assuming $a^5 b^2 = 1$, find the value of $\log_a a^4 b^5$.

$$a^5 b^2 = 1 \Rightarrow \log_a (a^5 b^2) = 5 + 2 \log_a b = 0 \Rightarrow \log_a b = -\tfrac{5}{2}.$$

$$\text{Thus, } \log_a (a^4 b^5) = 4 + 5 \log_a b = 4 + 5(-\tfrac{5}{2}) = \tfrac{8-25}{2} = -\tfrac{17}{2}.$$

If not quite sure of the idea behind the processes above, follow the steps below:

Solving a problem in math, we break the problem into pieces, and then, put the pieces together so that they add up to the solution.

Basically, this problem is made of a log, $\log_a a^4 b^5$ and an equality, $a^5 b^2 = 1$.
So we probably have to break into pieces the log and equality.

Breaking up $\log_a a^4 b^5$, we can get: $\log_a a^4 b^5 = \log_a a^4 + \log_a b^5 = 4 + 5 \log_a b$.
So we get: $\log_a a^4 b^5 = 4 + 5 \log_a b$.
Then, it seems we want to get the value of $\log_a b$. Where do we get the value, though?

We have: $a^5 b^2 = 1$ to work with.
So we probably can get the value of $\log_a b$ out of the equality given, $a^5 b^2 = 1$. How?

Applying $\log_a$ to $a^5 b^2 = 1$, we get: $\log_a a^5 b^2 = \log_a 1 = 0 \Rightarrow \log_a a^5 b^2 = 0$.

Next, we get: $\log_a a^5 b^2 = \log_a a^5 + \log_a b^2 = 5 + 2 \log_a b = 0 \Rightarrow \log_a b = -\tfrac{5}{2}$.

Now, going back to $\log_a a^4 b^5 = 4 + 5 \log_a b$, and do the substitution, we get:

$\log_a a^4 b^5 = 4 + 5 \log_a b = 4 + 5(-\tfrac{5}{2}) = \tfrac{8-25}{2} = -\tfrac{17}{2}$, which is the solution.

Suggestions or Solutions
To the Problem in the Example 2

Assuming $f(x) = 2^{\frac{x}{2}}$, **evaluate** $f(x)$ **at** $x = 2\log_2 3 + \log_4 6 - \log_8 2\sqrt{2}$.

First, we can get: $\log_4 6 = \log_{2^2} 6 = \frac{1}{2}\log_2 6 = \log_2 \sqrt{6}$, and

$\log_8 2\sqrt{2} = \log_{2^3} 2^{\frac{3}{2}} = \frac{1}{3} \cdot \frac{3}{2}\log_2 2 = \frac{1}{2}\log_2 2 = \log_2 \sqrt{2}$.

So next, we can get: $x = 2\log_2 3 + \log_4 6 - \log_8 2 = \log_2 3^2 + \log_2 \sqrt{6} - \log_2 \sqrt{2}$

$= \log_2 \frac{3^2 \sqrt{6}}{\sqrt{2}} = \log_2 3^2 \sqrt{3} = \log_2 3^{\frac{5}{2}} = \frac{5}{2}\log_2 3$.

Next, setting $f(\frac{5}{2}\log_2 3) = 2^a$, we get: $a = \frac{x}{2} = \frac{\frac{5}{2}\log_2 3}{2} = \frac{5}{4}\log_2 3 = \log_2 3^{\frac{5}{4}}$.

So next, setting $A = f(\frac{5}{2}\log_2 3)$, we get: $\log_2 A = \log_2 2^a = a = \log_2 3^{\frac{5}{4}} \Rightarrow A = 3^{\frac{5}{4}}$.

If not quite sure of the idea behind the processes above, follow the steps below:

We have a log identity where $\log_{b^m} A^n = \frac{n}{m}\log_b A$. How come?

We have a log identity where $\log_x y = \dfrac{\log_k y}{\log_k x}$.

So we can get: $\log_{b^m} A^n = n\log_{b^m} A = n\dfrac{\log_b A}{\log_b b^m} = n\dfrac{\log_b A}{m\log_b b} = n\dfrac{\log_b A}{m} = \frac{n}{m}\log_b A$.

For instance, we can have: $\log_8 9 = \log_{2^3} 3^2 = \frac{2}{3}\log_2 3$.

So to begin with, we can get:

$$\log_4 6 = \log_{2^2} 6 = \tfrac{1}{2}\log_2 6 = \log_2 \sqrt{6}.$$

$$\log_8 2\sqrt{2} = \log_{2^3} 2^{\frac{3}{2}} = \tfrac{1}{3}\cdot\tfrac{3}{2}\log_2 2 = \tfrac{1}{2}\log_2 2 = \log_2 \sqrt{2}.$$

So next, simplifying the value given, we get:

$$2\log_2 3 + \log_4 6 - \log_8 2 = \log_2 3^2 + \log_2 \sqrt{6} - \log_2 \sqrt{2}$$

$$= \log_2 \tfrac{3^2\sqrt{6}}{\sqrt{2}} = \log_2 3^2\sqrt{3} = \log_2 3^{\frac{5}{2}} = \tfrac{5}{2}\log_2 3.$$

Thus, $2\log_2 3 + \log_4 6 - \log_8 2\sqrt{2}$ is nothing but $\tfrac{5}{2}\log_2 3$.

So the solution is the value of $f(\tfrac{5}{2}\log_2 3)$.

Let's first, set: $f(\tfrac{5}{2}\log_2 3) = 2^a$.

Then, we get: $2^{\frac{x}{2}} = 2^a$, and thus, we get: $a = \tfrac{x}{2} = \tfrac{\frac{5}{2}\log_2 3}{2} = \tfrac{5}{4}\log_2 3 = \log_2 3^{\frac{5}{4}}$.

So we get: $a = \log_2 3^{\frac{5}{4}}$, and thus, can readily see that the solution is $3^{\frac{5}{4}}$ by another log identity where $Q = b^{\log_b Q}$. How come?

Assuming $x = b^{\log_b Q}$, and taking $\log_b$ of both sides, we get:

$$\log_b x = \log_b b^{\log_b Q} = (\log_b Q)(\log_b b) = \log_b Q \Rightarrow \log_b x = \log_b Q \Rightarrow x = Q.$$

What if however, we forget that identity?

We know: $a = \log_2 3^{\frac{5}{4}}$, so taking $\log_2$ of both sides of $f(\tfrac{5}{2}\log_2 3) = 2^a$, we get:

$$\log_2 f(\tfrac{5}{4}\log_2 3) = \log_2 2^a = a = \log_2 3^{\frac{5}{4}} \Rightarrow \log_2 f(\tfrac{5}{4}\log_2 3) = \log_2 3^{\frac{5}{4}} \Rightarrow f(\tfrac{5}{4}\log_2 3) = 3^{\frac{5}{4}}$$

Suggestions or Solutions
To the Problem in the Example 3

Show that $a^{\log b} = b^{\log a}$.

$$\log a^{\log b} = (\log b)(\log a) = (\log a)(\log b) = \log b^{\log a} \Rightarrow \log a^{\log b} = \log b^{\log a}$$

$$\Rightarrow a^{\log b} = b^{\log a}.$$

If not quite sure of the idea behind the processes above, follow the steps below:

When solving log equations or doing log algebra, if taking logs of both sides of the equation, equality, inequality, etc., we often see the passage through which we can get to the solution.

Now, we know that the equation given has **log b** and **log a**, which are common logs. So assuming now, $x = a^{\log b}$, and taking the common log of both sides, we get:

$$\log x = \log a^{\log b} = (\log b)(\log a) = (\log a)(\log b) = \log b^{\log a} \Rightarrow \log x = \log b^{\log a}$$

$$\Rightarrow x = b^{\log a}.$$

Thus, we now have: $x = a^{\log b}$, and $x = b^{\log a}$.

Therefore, we get: $a^{\log b} = b^{\log a}$.

Thus, we can swap the base and antilog when working with an expression in such a form as $m^{\log n}$, where n is the antilog in a common log, and m is the base.

So for instance, if both $(x^2 - y + 2)$ and $(2y + 3z - 1)$ are positive, we can get:

$$(x^2 - y + 2)^{\log(2y + 3z - 1)} = (2y + 3z - 1)^{\log(x^2 - y + 2)}.$$

For another simple instance, we have: $5^{\log 3} = 3^{\log 5}$.

Suggestions or Solutions
To the Problem in the Example 4

Assuming $b^x = u$, $b^y = v$, and $b^z = w$, express $\log_{uv^2w^3} u^3v^2w$ in terms of x, y, and z.

$b^x = u \Leftrightarrow x = \log_b u$, $b^y = v \Leftrightarrow y = \log_b v$, and $b^z = w \Leftrightarrow z = \log_b w$.

$$\log_{uv^2w^3} u^3v^2w = \frac{\log_b u^3v^2w}{\log_b uv^2w^3} = \frac{\log_b u^3 + \log_b v^2 + \log_b w}{\log_b u + \log_b v^2 + \log_b w^3} = \frac{3\log_b u + 2\log_b v + \log_b w}{\log_b u + 2\log_b v + 3\log_b w}$$

So we get: $\log_{uv^2w^3} u^3v^2w = \dfrac{3x + 2y + z}{x + 2y + 3z}$.

If not quite sure of the idea behind the processes above, follow the steps below:

We may want to see first, what x, y, and z are so that we can use them to express the log.

They are embedded in the assumption that $b^x = u$, $b^y = v$, and $b^z = w$. So we want to isolate them.   How though?

By the definition for logs, we can extract them the way as follows:

$b^x = u \Leftrightarrow x = \log_b u$, $b^y = v \Leftrightarrow y = \log_b v$, and $b^z = w \Leftrightarrow z = \log_b w$.

And next, we want to see how $\log_{uv^2w^3} u^3v^2w$ is made.

So we want to break apart the log given so that we can see how it can be made.
That is to say that we want to get the breakdown on the log so that we can put it in terms of x, y, and z.

Don't just look at x, y, and z themselves only though.

We want to look at these, too: $\log_b u$, $\log_b v$, and $\log_b w$, since we have:

$x = \log_b u$, $y = \log_b v$, and $z = \log_b w$.

So breaking up the log, we can break it the way below:

$$\log_{uv^2w^3} u^3v^2w = \frac{\log_b u^3v^2w}{\log_b uv^2w^3} = \frac{\log_b u^3 + \log_b v^2 + \log_b w}{\log_b u + \log_b v^2 + \log_b w^3} = \frac{3\log_b u + 2\log_b v + \log_b w}{\log_b u + 2\log_b v + 3\log_b w}$$

And we know: $x = \log_b u$, $y = \log_b v$, and $z = \log_b w$.

So we get: $\log_{uv^2w^3} u^3v^2w = \dfrac{3x+2y+z}{x+2y+3z}$.

So the key is the identity where $\log_p Q = \dfrac{\log_r Q}{\log_r p}$, and therefore, changing bases is quite similar to changing denominators in fraction arithmetic.

For instance, $\dfrac{3+8}{5} = \dfrac{\frac{3}{7}+\frac{8}{7}}{\frac{5}{7}} = \dfrac{\frac{3}{19}+\frac{8}{19}}{\frac{5}{19}} = \dfrac{\frac{3}{12.9}+\frac{8}{12.9}}{\frac{5}{12.9}} = \ldots$

And thus, we can put $\dfrac{3\log_b u + 2\log_b v + \log_b w}{\log_b u + 2\log_b v + 3\log_b w}$ the way below, too:

$$\frac{3\log_b u + 2\log_b v + \log_b w}{\log_b u + 2\log_b v + 3\log_b w} = \frac{3\frac{\log u}{\log b}+2\frac{\log v}{\log b}+\frac{\log w}{\log b}}{\frac{\log u}{\log b}+2\frac{\log v}{\log b}+3\frac{\log w}{\log b}} = \frac{3\log u + 2\log v + \log w}{\log u + 2\log v + 3\log w}.$$

In short:

$b^x = u \Leftrightarrow x = \log_b u$, $b^y = v \Leftrightarrow y = \log_b v$, and $b^z = w \Leftrightarrow z = \log_b w$.

$$\log_{uv^2w^3} u^3v^2w = \frac{\log_b u^3v^2w}{\log_b uv^2w^3} = \frac{\log_b u^3 + \log_b v^2 + \log_b w}{\log_b u + \log_b v^2 + \log_b w^3} = \frac{3\log_b u + 2\log_b v + \log_b w}{\log_b u + 2\log_b v + 3\log_b w}$$

So we get: $\log_{uv^2w^3} u^3v^2w = \dfrac{3x+2y+z}{x+2y+3z}$.

Suggestions or Solutions
To the Problem in the Example 5

Find the ratio $a : b : c$, assuming that a, b, and $c > 0$, but $c \neq 1$, and that:

$\log_b (2b + c) + \log_b (2b - c) = 2$

$\log_{a+b} c + \log_{a-b} c = (2 \log_{a+b} c)(\log_{a-b} c)$ **where $c \neq 1$**

To begin with, in the first expression, we get:

$\log_b (2b + c) + \log_b (2b - c) = \log_b \{(2b + c)(2b - c)\} = \log_b (4b^2 - c^2) = 2$
$\Rightarrow 4b^2 - c^2 = b^2 \Rightarrow 3b^2 = c^2$.

Next, on the left hand side in the second expression, we get:

$$\log_{a+b} c + \log_{a-b} c = \frac{1}{\log_c (a+b)} + \frac{1}{\log_c (a-b)} = \frac{\log_c (a-b) + \log_c (a+b)}{\{\log_c (a+b)\}\{\log_c (a-b)\}}$$

$$= \frac{\log_c (a^2 - b^2)}{\{\log_c (a+b)\}\{\log_c (a-b)\}}.$$

Next, on the right hand side, we get:

$$(2 \log_{a+b} c)(\log_{a-b} c) = \frac{2}{\log_c (a+b)} \cdot \frac{1}{\log_c (a-b)} = \frac{2}{\{\log_c (a+b)\}\{\log_c (a-b)\}}.$$

So we get $(2 \log_{a+b} c)(\log_{a-b} c) = (2 \log_{a+b} c)(\log_{a-b} c) \Rightarrow \log_c (a^2 - b^2) = 2 \Rightarrow a^2 - b^2 = c^2$.

Besides, we have: $3b^2 = c^2$.   Thus, $a^2 - b^2 = c^2 = 3b^2 \Rightarrow a^2 = 4b^2$.

And we know a, b, and c are all positive.
So we get: $a^2 = 4b^2 \Rightarrow a = 2b$, and $3b^2 = c^2 \Rightarrow c = \sqrt{3}b$.

Therefore, we can see that the ratio is $a : b : c = 2 : 1 : \sqrt{3}$.

If not quite sure of the idea behind the processes above, follow the steps below:

To begin with, simplifying the first equation, we get:

$$\log_b (2b + c) + \log_b (2b - c) = \log_b \{(2b + c)(2b - c)\} = \log_b (4b^2 - c^2) = 2.$$

Thus, we get: $\log_b (4b^2 - c^2) = 2$.

And by the definition for logs, we get: $4b^2 - c^2 = b^2$, so we get: $3b^2 = c^2$.

Next, moving on to $\log_{a+b} c + \log_{a-b} c = (2 \log_{a+b} c)(\log_{a-b} c)$, we have: $c \neq 1$.

So beginning with the left hand side, we can put it the way below:

$$\log_{a+b} c + \log_{a-b} c = \frac{1}{\log_c (a + b)} + \frac{1}{\log_c (a - b)}. \quad \text{How come?}$$

We have a log identity, $\log_x y = \dfrac{\log_y y}{\log_y x} = \dfrac{1}{\log_y x}$, where x and $y > 0$ but $\neq 1$.

And next, moving on to the right hand side, $(2 \log_{a+b} c)(\log_{a-b} c)$, we get:

$$(2 \log_{a+b} c)(\log_{a-b} c) = \frac{2}{\log_c (a + b)} \cdot \frac{1}{\log_c (a - b)} = \frac{2}{\{\log_c (a + b)\} \{\log_c (a - b)\}}.$$

So we get: $(2 \log_{a+b} c)(\log_{a-b} c) = \log_{a+b} c + \log_{a-b} c = \dfrac{1}{\log_c (a + b)} + \dfrac{1}{\log_c (a - b)}$

$$\Rightarrow \frac{1}{\log_c a + b} + \frac{1}{\log_c a - b} = \frac{2}{(\log_c a + b)(\log_c a - b)}.$$

Meanwhile:

$$\frac{1}{\log_c (a + b)} + \frac{1}{\log_c (a - b)} = \frac{\log_c (a - b) + \log_c (a + b)}{\{\log_c (a + b)\} \{\log_c (a - b)\}}, \text{ and}$$

$$\log_c (a - b) + \log_c (a + b) = \log_c \{(a - b)(a + b)\} = \log_c (a^2 - b^2).$$

Thus, we get: $\dfrac{1}{\log_c(a+b)}+\dfrac{1}{\log_c(a-b)}=\dfrac{\log_c(a^2-b^2)}{\{\log_c(a+b)\}\{\log_c(a-b)\}}$.

So we get:

$$\dfrac{\log_c(a^2-b^2)}{\{\log_c(a+b)\}\{\log_c(a-b)\}}=\dfrac{2}{\{\log_c a+b)\}\{\log_c(a-b)\}}\Rightarrow \log_c(a^2-b^2)=2\Rightarrow a^2-b^2=c^2.$$

Besides, we have: $3b^2 = c^2$. Thus, we get: $a^2 - b^2 = c^2 = 3b^2 \Rightarrow a^2 = 4b^2$.

And we know *a, b,* and *c* are all positive.

So we get: $a^2 = 4b^2 \Rightarrow a = 2b$, and $3b^2 = c^2 \Rightarrow c = \sqrt{3}b$.

Therefore, we can see that the ratio is $a:b:c = 2:1:\sqrt{3}$.

Note:

The ratio above indicates relative values and not actual values.

So *a* is twice *b*, and *c* is $\sqrt{3}$ of *b*.

Thus, if the actual value of *b* is *k*, then that of *a* is *2k*, and that of *c* is $\sqrt{3}k$ where $k \neq 1$.
How come $k \neq 1$?

That's only because of the nature of logarithms.

The value of *b* itself cannot be 1 because it is used as a base in a logarithm, and thus, is not allowed to be 1.

In the problem definiton, *c* is not allowed to be 1, either.

Thus, *b* cannot be $\frac{1}{\sqrt{3}}$, either, in this particular case.

In short:

To begin with, in the first expression, we get:

$\log_b (2b + c) + \log_b (2b - c) = \log_b \{(2b + c)(2b - c)\} = \log_b (4b^2 - c^2) = 2$
$\Rightarrow 4b^2 - c^2 = b^2 \Rightarrow 3b^2 = c^2.$

Next, on the left hand side in the second expression, we get:

$$\log_{a+b} c + \log_{a-b} c = \frac{1}{\log_c (a+b)} + \frac{1}{\log_c (a-b)} = \frac{\log_c (a-b) + \log_c (a+b)}{\{\log_c (a+b)\}\{\log_c (a-b)\}}$$

$$= \frac{\log_c (a^2 - b^2)}{\{\log_c (a+b)\}\{\log_c (a-b)\}}.$$

Next, on the right hand side, we get:

$$(2\log_{a+b} c)(\log_{a-b} c) = \frac{2}{\log_c (a+b)} \cdot \frac{1}{\log_c (a-b)} = \frac{2}{\{\log_c (a+b)\}\{\log_c (a-b)\}}.$$

So we get $(2 \log_{a+b} c)(\log_{a-b} c) = (2 \log_{a+b} c)(\log_{a-b} c) \Rightarrow \log_c (a^2 - b^2) = 2 \Rightarrow a^2 - b^2 = c^2.$

Besides, we have: $3b^2 = c^2.$

Thus, $a^2 - b^2 = c^2 = 3b^2 \Rightarrow a^2 = 4b^2.$

And we know **a, b,** and **c** are all positive.

So we get: $a^2 = 4b^2 \Rightarrow a = 2b$, and $3b^2 = c^2 \Rightarrow c = \sqrt{3}b.$

Therefore, we can see that the ratio is $a : b : c = 2 : 1 : \sqrt{3}.$

A. Common Logs

To begin with, what kind of number can we take a log of?

If a number is positive, we can take a log of it.
Next, a log has its base, which is a number. What number then, can we use as the base?

Taking a log, we can use as the base any number positive and unequal to 1.
Quite usually though, when taking logs, we use 10 as the base. And such a log is called a **common log**. That's because it is commonly used, that is, it is used quite often.

So taking a log of a number to base 10, or taking a log sub 10 of a number, we take a *common log* of the number. And taking a common log, we use 10 as the base.

When working with numbers very large or small, we often find common logs are convenient and effective. Taking a common log of a number, we can readily see how big or small the numbers is.   How can we see it though?

The value of a common log is specified by two parts.
One is called a **characteristic**, and the other is called a **mantissa**.

And taking the value of the common log of a number, and finding the characteristic, we can quickly see the highest place value of the number. What is the highest place value?

It is the place value of the highest digit. So for instance, the highest place value of 5023 is $1000 = 10^3$, and the highest place value of 0.04082 is $0.01 = 10^{-2}$. That is, in 5023, 5 is in the 1000's place, and in 0.04082, 4 is in the one hundredth's place.

And looking at the characteristic, we can readily see the highest place value.
How then, can we get or identify the characteristic of a common log?

Math begins with a definition. So let's now, begin with the definition for common logs.

Normally, taking a common log of a number, we omit the base 10. Thus, seeing a log of a number to no base, we can just take it as the common log of the number.

So the definition for common logs can be as follows: $y = \log x \Leftrightarrow x = 10^y$, where $x > 0$.

That is to say that if $x > 0$, $y = \log x \Rightarrow x = 10^y$, and also, $x = 10^y \Rightarrow y = \log x$, too.
So $\log x$ is the same as $\log_{10} x$, and is taken as a common log.

For instance, $2 = \log 100 \Rightarrow 100 = 10^2$, and also, $100 = 10^2 \Rightarrow 2 = \log 100$, too.
So $\log_{10} 100 = \log 100$, which is taken as a common log.

How then, can we get or identify the characteristic of a common log?

Suppose $N > 0$.
Then, we can put the value of the common log of N in terms of the characteristic and mantissa the way as follows: $\log_{10} N = c + m$, where c is an *integer*, and $0 \le m < 1$.

And in the equality above, c is called a *characteristic*, and m is called a *mantissa*.

Normally though, taking a common log of N, we don't specify the base, so we put it this way: **log N**.

And set it this way: **log $N = c + m$**, where c is an *integer*, and $0 \leq m < 1$.

For instance, if **log $A = 3.1405$**, the characteristic is 3, and the mantissa is 0.1405.

So what's good about a common log?

First, the integer c called characteristic can quickly tell us the highest place value of N. And the highest place value of N is simply 10^c.

So for instance, taking a common log of 3851, we can set it the way below:

log $3851 = 3 + m$ where $0 \leq m < 1$.

So the characteristic $c = 3$ can tell us immediately the highest place value is $10^c = 10^3$. And the same is true, too, for any number where the highest place value is 10^3.

Thus for instance, taking a common log of 9759.024, we can set it the way below:

log $9759.024 = 3 + n$ where $0 \leq n < 1$.

And by the same token, we can set: **log $8076.1205 = 3 + p$** where $0 \leq p < 1$.

So if the highest place value of every antilog is the same, only the mantissa is different.

And next, if for instance, we take common logs of 305, 3050, and 30500, we get the same mantissas. That is to say that we can set:

log $305 = 2 + m$, log $3050 = 3 + m$, and **log $30500 = 4 + m$**, where $0 \leq m < 1$.

Taking care of the characteristic though, we need to pay attention to the sign of it.
In other words, the sign of the characteristic matters.
And the sign depends on the *magnitude* of the antilog.

For instance, assuming $\log N = c + m$, where c is an integer, and $0 \le m < 1$, we can get:

If N is ≥ 1, we get: $c \ge 0$.
If N is greater than 0 but less than 1, that is, $0 < N < 1$, we get c negative.

In sum: $N \ge 1 \Rightarrow c \ge 0$, and $0 < N < 1 \Rightarrow c < 0$.

Next, in the case where $\log N = c + m$, $(c + m)$ is a log value, which is the value of $\log N$.
When getting the value of the $\log N$ though, how do we get the characteristic c?

If the antilog N has n digits above the decimal point, we get: $c = n - 1$.

In other words, N has $c + 1$ digits above the point, since $n = c + 1$.

For instance, if $N = 3809.07042$, we get: $c = 3$, since N has 4 digits above the point.

And it doesn't matter how many digits N has below the point.
What matters is *the number of digits above the point*.
So for instance, the characteristic of $\log 102.31$ is 2, and so is that of $\log 927.993527$.

On the other hand, if N has no nonzero digit above the decimal point as 0.01, and its first nonzero digit appears in the nth place below the point, then we get: $c = -n$.

For instance, if the antilog $N = 0.004208$, we get: $c = -3$, since the first nonzero digit appears in the third place below the point.

Anyhow, the highest place value of the antilog N is 10^c, where c is the characteristic.

In the examples above:
$N = 3809.07042 \Rightarrow c = 3 \Rightarrow 10^c = 10^3$, so the highest place value of N is 1000.
$N = 0.004208 \Rightarrow c = -3 \Rightarrow 10^c = 10^{-3} = 0.001$, so the highest place value of N is 0.001.

So putting threads together, what does the characteristic have to do with?

It has to do with the magnitude of the highest digit in the number we take a common log of. And in fact, it can tell us such a magnitude, which is 10^c, where c is the characteristic. That is, it says the highest place value of the antilog in a common log.

- Now, what about the **mantissa**?

Suppose two positive numbers have the same sequence of digits, but the positions of the decimal point are different.

For instance, if $x = 1202.387$ and $y = 0.01202387$, then x and y have the same sequence of digits, but the positions of the decimal point are different.

Then, if we take common logs of x and y above, the characteristics of the logs will be different, but the mantissas will be the same.

That is, taking the difference between **log x** and **log y**, we get an integer.

The highest place value of x above is 10^3, so the characteristic c of **log x** is 3.

The highest place value of y above is 10^{-2}, and therefore, c of **log y** is -2.

So the characteristics are different, because the highest place values are different.

However, the mantissas will be the same, because the sequences of digits are the same.

Thus, we can set: **log x = 3 + m**, and **log y = -2 + m**, where $0 \le m < 1$.

Therefore, we get: **log x – log y = 5**, which is an integer, of course.

So what does the mantissa have to do with?

It has to do with the sequence of digits in the number we take a common log of. And in fact, if two numbers have the same sequence of digits as 1023 and 1.023, and we take common logs of the two numbers, their mantissas are the same.

Let's next, have a look at cases where $0 < N < 1$, that is, the antilog is between 0 and 1.

So suppose $\log N = c + m$, where $0 < N < 1$, c is an integer, and $0 \leq m < 1$.

Then, the characteristic c is negative.

That is, if the antilog in a common log is between 0 and 1, the characteristic c is negative. In such a case, we use special notation for the characteristic c of $\log N$.

In the special notation, we place the minus sign (–) on top of the characteristic.

So for instance, we normally set: $\log y = \bar{2}.abcd$ instead of $\log y = -2.abcd$.

Thus, unlike cases of ordinary numbers, we put a minus sign not in front but on top of the characteristic.

Why is the characteristic c negative though, if $0 < N < 1$, and also, why do we bother putting the minus sign above c?

The answer to the first part of the question begins with a statement that the highest place value of a number between 0 and 1 is less than or equal to 10^{-1}, which is one tenth.

So of the common log of such a number, the characteristic is ≤ -1 since the highest place value of the number is 10^c, where c is the characteristic of the common log.

Suppose for instance, that the antilog $N = 0.3$. Then, we can get: $N = 0.3 = 0.1 \cdot 3 = 10^{-1} \cdot 3$.

So we can get: $\log N = \log 10^{-1} \cdot 3 = \log 10^{-1} + \log 3 = -1 + \log 3$.

Thus, we get: $\log 0.3 = -1 + \log 3$.

And we know: **log 1 = 0** and **log 10 = 1**.

So we get: $1 \leq 3 < 10 \Rightarrow$ **log 1 ≤ log 3 < log 10 ⇒ 0 ≤ log 3 < 1**.

Thus, we get: **log 0.3 = -1 + s,** where **0 ≤ s < 1**.

So the characteristic is -1, which is negative.

Suppose for another instance, that $N = 0.999$. Then, $N = 0.999 = 0.1 \cdot 9.99 = 10^{-1} \cdot 9.99$.

So **log** N = **log 0.999 = log** 10^{-1} + **log 9.99 = -1 + log 9.99** $\Rightarrow$ **log 0.999 = -1 + log 9.99**.

And since **log 1 = 0** and **log 10 = 1**, we get: $1 \leq 9.99 < 10 \Rightarrow 0 \leq$ **log 9.99 < 1**.

Therefore, **log 0.999 = -1 + t**, where **0 ≤ t < 1**.   What if $N = 0.0299$?

Then, we can get: $N = 0.0299 = 0.01 \cdot 2.99 = 10^{-2} \cdot 2.99$.

So we can get: **log** N = **log 0.0299 = log** 10^{-2} + **log 2.99 = -2 + log 2.99**.

And since **log 1 = 0** and **log 10 = 1**, we get: $1 \leq 2.99 < 10 \Rightarrow 0 \leq$ **log 2.99 < 1**.

Therefore, **log 0.0299 = -2 + u** where **0 ≤ u < 1**.

Now, going back to the second part of the question, we have:

If **0 < N < 1**, why do we bother putting the minus sign above *c*?

The answer begins with a statement that the value of a common log is not just a value but a string of digits showing two values separated by the decimal point, one value is called a characteristic, and the other is called a mantissa.

So we need to differentiate them first.

And next, if we put the negative sign in front, the mantissa is negative, too, which is against the definition where **0 ≤ the mantissa < 1**.

Now in a common log, **log N**, if **0 < N < 1**, then we get: **log N < 0**.

So going back to the examples above, we have:

log 0.3 = -1 + s where **0 ≤ s < 1**, so **log 0.3 < 0** since **-1 + s < 0** since **0 ≤ s < 1**.

log 0.999 = -1 + t where **0 ≤ t < 1**, so **log 0.999 < 0** since **-1 + t < 0** since **0 ≤ t < 1**.

log 0.0299 = -2 + u where **0 ≤ u < 1**, so **log 0.0299 < 0** since **-2 + u < 0** since **0 ≤ u < 1**.

Suppose for another instance, we set: **log N = -1.2345**.

Then, the characteristic is not -1, and the mantissa is not 0.2345, either.

That's because we have: **0 ≤ the mantissa < 1**, and **-1.2345 = -1 − 0.2345**.

In fact, the mantissa of **log N** is **1 − 0.2345 = 0.7655**, and the characteristic is **-2**.

That's because **log N = -1.2345 = -1 − 1 + 1 − 0.2345 = -2 + 0.7655 = $\overline{2}.7655$**.

Suppose for another instance, **log N = $\overline{1}.3582$**.

Then, we get: **log N = -1 + 0.3582**, so it shows the characteristic and mantissa properly.

So if N in **log N** is between 0 and 1, then:

We need to note that the characteristic is negative, but the **mantissa** is **positive**, which is the nature of a common log, and the nature doesn't change no matter what positive value the antilog N may have. In other words, we want to keep the mantissa positive.

Thus, putting the minus sign in front of the characteristic, we cannot keep the mantissa positive. So such an unusual placement of the minus sign is the means for that purpose.

Suppose for another instance, **log N = -1.1812**, and **log M = $\overline{1}.1812$**. Then, we get:

log N = -1.1812 ⇒ log N = -1 + (-0.1812) = -1 − 0.1812.

log M = $\overline{1}.1812$ ⇒ log M = -1 + 0.1812.

Note:

Using a calculator, and taking the common log of a number between 0 and 1, we just get a negative number in the display panel.

For instance, using a calculator, we get: log 0.02 ≈ -1.69897 instead of $\overline{2}$.30103.

That's because -1.69897 = -2 + 0.30103, where -2 is the characteristic, and 0.30103 is the mantissa.

Now, reconsidering those *x* and *y* taken for example above, we have:

x = **1202.387**, and *y* = **0.01202387**.

And taking common logs of *x* and *y*, we get: **log *x* = 3.*abcd***, and **log *y* = $\overline{2}$.*abcd***.

More specifically:

log *x* = log 1000(1.202387) = log 10^3 + log 1.202387 = 3 + log 1.202387
log *y* = log 0.01(1.202387) = log 10^{-2} + log 1.202387 = -2 + log 1.202387

So the mantissas of both logs are the same, so for instance, both are **0.*abcd***.
Thus, the mantissa of **log *y*** is not negative. Neither is that of **log *x***, of course.

So **log *y*** is not -2 + (-0.*abcd*) but **-2 + 0.*abcd***.

Now, if *N* ≥ **1**, why does the characteristic *c* of **log *N*** have to be ≥ **0**?

That's because the highest place value of a number ≥ 1 has to be ≥ 10^0.

So of the common log of such a number, the characteristic is ≥ **0** since the highest place value of a number is 10^c where *c* is the characteristic of the common log of the number. Still not quite sure?

Let's take a look at some examples that can help us see the reason better.

Suppose for instance, that the antilog $N = 1$.   Then, $N = 1 = 10^0$.

So $\log N = \log 10^0 = 0 \log_{10} 10 = 0 \cdot 1 = 0$.

We have: $\log N = c + m$ where c is an integer, and $0 \leq m < 1$. Thus, $c = 0$ and $m = 0$.

So anyway, it is true that $c \geq 0$, since $c = 0$.

Suppose for another instance, that $N = 3$. Then, $N = 3 = 1 \cdot 3 = 10^0(3)$.

Thus, $\log N = \log 3 = \log 10^0 + \log 3 = 0 + \log 3$. So $\log N = 0 + \log 3$.

And since $\log 1 = 0$ and $\log 10 = 1$, we get: $1 \leq 3 < 10 \Rightarrow 0 \leq \log 3 < 1$.

Thus, $\log 3 = 0 + v$ where $0 \leq v < 1$.

And we have: $\log N = c + m$ where c is an integer, and $0 \leq m < 1$.

Therefore, $c = 0$ and $m = v$.   So anyway, it is true that $c \geq 0$ since $c = 0$.

Suppose for one more instance, that $N = 358$.   Then, $N = 358 = 100(3.58) = 10^2(3.58)$.

Thus, $\log N = \log 358 = \log 10^2 + \log 3.58 = 2 + \log 3.58$.

And since $\log 1 = 0$ and $\log 10 = 1$, we get: $1 \leq 3.58 < 10 \Rightarrow 0 \leq \log 3.58 < 1$.

Thus, $\log 358 = 2 + w$ where $0 \leq w < 1$.

And we have: $\log N = c + m$ where c is an integer, and $0 \leq m < 1$.

Therefore, $c = 2$ and $m = w$.   So anyway, it is true that $c \geq 0$, since $c = 2$.

Besides a common log, we have another frequent log, where the base is normally not specified either.

Such a log is called a natural log, which is quite popular with many scientists.

In a natural log, the base is *e*, called Euler's number, and we can see it quite often taking courses in calculus.

Conceptually, the number *e* can be described as $(1 + \boldsymbol{0})^\infty$ where $\boldsymbol{0}$ is called infinitesimal, and ∞ is called infinity. It is in fact, an irrational number, and can be roughly estimated to be 2.72. To be more accurate, it is approximately 2.718181828459045.

And taking a natural log of a number, we can take the log of the number to base *e*.
That is, we can take log sub *e* of the number.
So for instance, taking the natural log of 3, we get **$\log_e 3$**.

However, we often use a different sign, which is **ln**. And we use it without the base *e*.

So for instance, taking the natural log of 3, we get **ln 3**.

And thus, we have: **$\log_e 3$ = ln 3**, which can be read as natural log of 3 or just LN of 3.

In addition, we often use a capital **L**, too, instead of the lower case **l**.

So we have: **Ln N = ln N = $\log_e N$**.

Anyway, taking a common log of *A*, we can set: **$\log A = c + m$**, where *c* is an integer, and **$0 \le m < 1$**. And we call *c* the characteristic, and call *m* the mantissa.

How come can we set it that way, though?

We can see it doing the examples in common logs.

Examples 1 in Common Logs

Note that a small dot · is a multiplication operator unless specified otherwise, e.g. $3 \times 5 = 3 \cdot 5 = 15$.

Note also, that no use of calculator is allowed in this set of examples.

0. Assuming $\log X = c + m = -1.123$, find the characteristic c and the mantissa m.

1. Assuming $\log Y = c + m = -0.207$, find the characteristic c and the mantissa m.

2. Put the expressions below into the form of $\bar{2}.309$.

2.0. $\bar{1}.4256 + \bar{3}.7233$

2.1. $\bar{3}.4263 - \bar{2}.2573$

2.2. $\bar{2}.3751 \cdot 4$

2.3. $\dfrac{\bar{3}.2564}{2}$

Suggestions or Solutions
To the Problems in the Examples 1

0. Assuming $\log X = c + m = -1.123$, find the characteristic c and the mantissa m.

Let's briefly check the basics of common logs.

The **base in a common log** is **10**, which is usually omitted. So assuming $K > 0$, and taking the common log of K, we get: $\log K = c + m$, where c is an integer called the characteristic, m is called the mantissa, and $0 \leq m < 1$. Again, $0 \leq$ **the mantissa** < 1.

Now, we have: $\log X = c + m = -1.123 = -1 - 0.123$.
However, the mantissa is not -0.123, because $0 \leq$ (the mantissa m) < 1.

So we want to extract the mantissa m from **-0.123**.
$-0.123 = -1 + 1 - 0.123 = -1 + 0.877 \Rightarrow -0.123 = -1 + 0.877$.

So the mantissa m is 0.877.

Thus, $-1.123 = -1 - 0.123 = -1 - 1 + 0.877 = -2 + 0.877 = \overline{2}.877$.

Therefore, $c = -2$ and $m = 0.877$.

In short:

$-1.123 = -1 - 0.123 = -1 - 1 + 1 - 0.123 = -1 - 1 + 0.877 = -2 + 0.877 = \overline{2}.877$.
$\therefore c = -2$, and $m = 0.877$.

Let's now, consider a general case.

Suppose $\log K = -A.BCD$, where **A, B, C,** and **D** are single digit integers.

Then, we can put it this way: $\log K = -A - 0.BCD$.
(For instance, if $\log K = -2.1093$, we get: $\log K = -2 - 0.1093$.)

We know however: $0 \leq$ **the mantissa** < 1.

So **-0.BCD** is not the mantissa, and **-A** is not the characteristic, either.

And we know: $1 < 1 - 0.BCD < 0$.

So we can put it this way: $\log K = -A - 0.BCD = -A - 1 + 1 - 0.BCD$.

Therefore, the characteristic c is: $-A - 1 = -(A + 1)$, and the mantissa m is: $1 - 0.BCD$.

On the other hand, if $\log K = \overline{A}.BCD$, we get: $\log K = -A + 0.BCD$.

Therefore, the characteristic c is **-A**, and the mantissa m is **0.BCD**.

1.   Assuming $\log Y = c + m = -0.207$, find the characteristic c and the mantissa m.

The mantissa is not -0.207, because $0 \leq$ (the mantissa m) < 1.

So we want to extract the mantissa m from **-0.207**.
$-0.207 = -1 + 1 - 0.207 = -1 + 0.793 \Rightarrow -0.207 = -1 + 0.793 = \overline{1}.793$.

Therefore, the mantissa m is 0.793, and the characteristic c is -1.

In short:

$-0.207 = -1 + 1 - 0.207 = -1 + 0.793 = \overline{1}.793$.
$\therefore c = -1$, and $m = 0.793$.

112

2.

To begin with, note that a mantissa cannot be negative, and thus, has to be ≥ 0.

If **log X = -1.1234**, the characteristic is not -1, and the mantissa is not 0.1234 either.

That's because **log X = -1 – 0.1234 = -1 – 1 + 1 – 0.1234 = -2 + 0.8766**.

So the characteristic is -2, and the mantissa is 0.8766.   If however, **log $Y = \overline{1}.1234$**, then:

log Y = -1 + 0.1234, so the characteristic is -1, and the mantissa is 0.1234.

1.0. $\overline{1}.4256 + \overline{3}.7233$
First, we can set: $\overline{1}.4256 = -1 + 0.4256$, and $\overline{3}74233 = -3 + 0.7233$.
Thus next, $\overline{1}.4256 + \overline{3}.7233$ = -1 + 0.4256 – 3 + 0.7233 = -4 + 1.1489 = -4 + 1 + 0.1489
= -3 + 0.1489 = $\overline{3}.1489$.

1.1. $\overline{3}.4263 - \overline{2}.2573$
We can set: $\overline{3}.4263 = -3 + 0.4263$, and $\overline{2}.2573 = -2 + 0.2573$.
So we get: $\overline{3}.4263 - \overline{2}.2573$ = -3 + 0.4263 – (-2 + 0.2573) = -3 + 0.4263 + 2 – 0.2573
= -1 + 0.1690 = $\overline{1}.1690$.

1.2. $\overline{2}.3751 \cdot 4$
We can set: $\overline{2}.3751 = -2 + 0.3751$.
So we get: $\overline{2}.3751 \cdot 4$ = (-2 + 0.3751)4 = -8 + 1.5004 = -8 + 1 + 0.5004
= -7 + 0.5004 = $\overline{7}.5004$.

1.3. $\dfrac{\overline{3}.2564}{2}$

We can set: $\overline{3}.2564 = -3 + 0.2564$.

Thus, we get: $\dfrac{\overline{3}.2564}{2} = \dfrac{-3 + 0.2564}{2} = -1.5 + 0.1282 = -1 - 0.5 + 0.1282$

$= -1 - 1 + 1 - 0.5 + 0.1282 = -2 + 0.5 + 0.1282 = -2 + 0.6282 = \overline{2}.6282$.

Examples 2 in Common Logs

Note that a small dot · is a multiplication operator unless specified otherwise, e.g. $3 \times 5 = 3 \cdot 5 = 15$.

Note also, that no use of calculator is allowed in this set of examples.

Use **log 1.35 = 0.1303**, and find the characteristic and mantissa of the common log of each of the numbers as follows.

0.   **13.5**

1.   **135**

2.   **0.135**

3.   **0.0135**

4.   **1.35^2**

5.   **$\sqrt{13.5}$**

6.   **0.135^{-2}**

7. $\dfrac{1}{\sqrt{0.0135}}$

Suggestions or Solutions
To the Problems in the Examples 2

Suppose that two different positive numbers have the same sequence of digits.

Then, the common logs of the two numbers have different characteristics, but their mantissas are the same.

The first four numbers given in this set of examples have the same sequence of digits, and the sequence has 1, 3, and 5 in order. Therefore, their mantissas are the same.

What then, about the next four numbers?

Those four numbers are powers and radicals, and their bases have the same sequence of digits, but if we expand those four numbers, the expansions will probably have different sequences of digits.

For instance, if we expand 2^3 and $\sqrt{2}$, the expansions are 8 and 1.4142135...

And also, the last four numbers in this set of examples have exponents other than 1. Technically, we can take the first four numbers for antilogs in power notation, too. For instance, 13.5 can be taken for an antilog where the base is 13.5, and the exponent is 1.

Now in this example, we use **log 1.35 = 0.1303** doing each problem.

So we may want to first, take a common log of each number given in each problem and then, extract the log of 1.35. Thus, we get to use some identities on logarithms, and they are as follows:

Assuming M, N, and $b > 0$, but $b \neq 1$, we have:

$\log_b b = 1$.

$\log_b MN = \log_b N + \log_b M$.

$A = \log_b N \Rightarrow PA = P \log_b N = \log_b N^P$.

0.   13.5

We know: 13.5 = 10·1.35. Taking the common log of it, we get: $\log_{10} 10 \cdot 1.35$, which is normally though, put this way: **log 10·1.35**. So the base 10 is usually omitted.

And we have: $\log_b b = 1$, and $\log_b MN = \log_b N + \log_b M$.

So we get: **log 13.5 = log 10 + log 1.35 = 1 + 0.1303**.
Therefore, the characteristic is 1, and the mantissa is 0.1303.

1.   135

We know: $135 = 10^2(1.35)$. Taking the common log of it, we get: **log 10^2(1.35)**.
And we have: $\log_b b = 1$, $\log_b MN = \log_b N + \log_b M$, and $P \log_b N = \log_b N^P$.

So we get: **log 13.5 = log 10^2 + log 1.35 = 2 log 10 + log 1.35 = 2 + 0.1303**.
Therefore, the characteristic is 2, and the mantissa is 0.1303.

2.   0.135

We know: $0.135 = 10^{-1}(1.35)$. Taking the common log of it, we get: **log 10^{-1}(1.35)**.
And we have: $\log_b b = 1$, $\log_b MN = \log_b N + \log_b M$, and $P \log_b N = \log_b N^P$.

So we get: **log 0.135 = log 10^{-1} + log 1.35 = (-1)log 10 + log 1.35 = -1 + 0.1303**.
Therefore, the characteristic is -1, and the mantissa is 0.1303.

3.   0.0135

We know: $0.0135 = 10^{-2}(1.35)$. Taking the common log of it, we get: **log 10^{-2}(1.35)**.
And we have: $\log_b b = 1$, $\log_b MN = \log_b N + \log_b M$, and $P \log_b N = \log_b N^P$.

So we get: **log 0.0135 = log 10^{-2} + log 1.35 = (-2)log 10 + log 1.35 = -2 + 0.1303**.
Therefore, the characteristic is -2, and the mantissa is 0.1303.

4.   1.35^2

We have: $1.35^2 = (1.35)^2$. Taking the common log of it, we get: $\log (1.35)^2$.

And we have: $\log_b b = 1$, $\log_b MN = \log_b N + \log_b M$, and $P \log_b N = \log_b N^P$.

So we get: $\log (1.35)^2 = 2 \log 1.35 = 2(0.1303) = 0.2606$.

Therefore, the characteristic is 0, and the mantissa is 0.2606.

5. $\sqrt{13.5}$

We have: $\sqrt{13.5} = (10 \cdot 1.35)^{\frac{1}{2}}$. Taking the common log of it, we get: $\log(10 \cdot 1.35)^{\frac{1}{2}}$.

And we have: $\log_b b = 1$, $\log_b MN = \log_b N + \log_b M$, and $P \log_b N = \log_b N^P$.

So we get: $\log(10 \cdot 1.35)^{\frac{1}{2}} = \frac{1}{2}\log(10 \cdot 1.35) = \frac{1}{2}(1 + 0.1303) = 0.5 + 0.06515 = 0.56515$.

Therefore, the characteristic is 0, and the mantissa is 0.56515.

6.   0.135^{-2}

We have: $(0.135)^{-2} = \{10^{-1}(1.35)\}^{-2}$.

Taking the common log of it, we get: $\log \{10^{-1}(1.35)\}^{-2}$.

And we have: $\log_b b = 1$, $\log_b MN = \log_b N + \log_b M$, and $P \log_b N = \log_b N^P$.

So we get: $\log \{10^{-1}(1.35)\}^{-2} = (-2)\log 10^{-1}(1.35) = -2(-1 + 0.1303)$

$= 2 - 0.2606 = 2 - 1 + 1 - 0.2606 = 1 + 0.7394$.

Therefore, the characteristic is 1, and the mantissa is 0.7394.

7. $\frac{1}{\sqrt{0.0135}}$

We have: $\frac{1}{\sqrt{0.0135}} = (10^{-2} \cdot 1.35)^{-\frac{1}{2}}$.

Taking the common log of it, we get: $\log(10^{-2} \cdot 1.35)^{-\frac{1}{2}}$.

And we have: $\log_b b = 1$, $\log_b MN = \log_b N + \log_b M$, and $P \log_b N = \log_b N^P$.

So we get: $\log(10^{-2} \cdot 1.35)^{-\frac{1}{2}} = -\frac{1}{2}\log(10^{-2} \cdot 1.35) = -\frac{1}{2}(-2 + 0.1303)$

$= 1 - 0.06515 = 0.93485$.

Therefore, the characteristic is 0, and the mantissa is 0.93485.

In short:

0.   log 13.5 = log 10(1.35) = log 10 + log 1.35 = 1 + log 1.35 = 1 + 0.1303 = 1.1303.

Therefore, the characteristic = 1, and the mantissa = 0.1303.

1.   log 135 = log $10^2$1.35 = log 10^2 + log 1.35 = 2 + log 1.35 = 2 + 0.1303 = 2.1303.

Therefore, the characteristic = 2, and the mantissa = 0.1303.

2.   log 0.135 = log 10^{-1}1.35 = log 10^{-1} + log 1.35 = -1 + log 1.35 = -1 + 0.1303.

Therefore, the characteristic = -1, and the mantissa = 0.1303.

3.   log 0.0135 = log 10^{-2}1.35 = log 10^{-2} + log 1.35 = -2 + log 1.35 = -2 + 0.1303.

Therefore, the characteristic = -2, and the mantissa = 0.1303.

4.   log 1.35^2 = 2 log (1.35) = 2(0.1303) = 0.2606.

Therefore, the characteristic = 0, and the mantissa = 0.2606.

5.   log $\sqrt{13.5}$ = log$(10\cdot1.35)^{\frac{1}{2}}$ = $\frac{1}{2}$log$(10\cdot1.35)$ = $\frac{1}{2}(1+0.1303)$ = 0.5 + 0.06515

= 0.56515.

Therefore, the characteristic is 0, and the mantissa is 0.5652.

6.   log 0.135^{-2} = log $\{10^{-1}(1.35)\}^{-2}$ = (-2)log $10^{-1}(1.35)$ = -2(-1 + 0.1303) = 2 – 0.2606

= 2 – 1 + 1 – 0.2606 = 1 + 0.7394.

Therefore, the characteristic is 1, and the mantissa is 0.7394.

7.   log $\frac{1}{\sqrt{0.0135}}$ = log$(10^{-2}\cdot1.35)^{-\frac{1}{2}}$ = $-\frac{1}{2}$log$(10^{-2}\cdot1.35)$ = $-\frac{1}{2}(-2+0.1303)$

= 1 – 0.06515 = 0.93485.

Therefore, the characteristic is 0, and the mantissa is 0.9348.

Examples 3 in Common Logs

Note that a small dot· is a multiplication operator unless specified otherwise, e.g. 3 × 5 = 3·5 = 15.

Note also, that no use of calculator is allowed in this set of examples.

3. Using **log 412 = 2.6149**, find x in each case below.

3.0. **log x = 3.6149**

3.1. **log x = $\overline{1}$.6149**

3.2. **log x = $\overline{2}$.6149**

3.3. **log 41.2 = x**

3.4. **log $0.412^2 = x$**

3.5. **log $\dfrac{1}{\sqrt{0.0412}} = x$**

3.6. **log 412^x = 1.3542**

3.7. **log 412^{2x} = $\overline{2}$.3462**

Suggestions or Solutions
To the Problems in the Examples 3

Let's begin with the basics of common logs.

Suppose first, K is a number positive.
Then, the common log of K is: $\log K = c + m$ where c is an integer called the characteristic indicating the highest place value of K, m is the mantissa, and $0 \le m < 1$. And the highest place value of K is 10^c.

Suppose next, two different positive numbers have the same sequence of digits. For instance, one is 1.4809, and the other is 148.09.

Then, the common logs of the two numbers have different characteristics, but their mantissas are the same. That is to say that, the log values share the same mantissa, but their characteristics are different.

Now in this example, we use: $\log 412 = 2.6149$, which is the value of a common log.

So in the log value 2.6149, 2 is the characteristic, and thus, indicates that the highest place value of the antilog is 10^2. The antilog is 412, and the common log of it is $\log 412$, of course. So if the common log of x is 0.6149, x is 4.12. That's for two reasons below:

One is that the characteristic is 0, so the highest place value is 10^0, which is 1.

And the other is that the mantissa is the same as that of $\log 412$, so the sequence of digits has 4, 1, and 2 in order. What then, if the common log of x is 1.6149?

Then, x is 41.2 because the characteristic is 1, so the highest place value is 10^1, and since the mantissa is the same as that of $\log 412$, the sequence of digits has 4, 1, and 2 in order.

Now, the first three logs given in this set of examples have the same mantissa, which is 0.6149.

In each of the first three examples, therefore, x needs to have the same sequence of digits, which has 4, 1, and 2 in order, since the mantissa is the same as that of **log 412**.

So using the characteristic in each of the three logs, we can get the value of x.

Let's next, move on to the forth example: **log 41.2 = x**.

The antilog 41.2 has the same sequence of digits as the one in 412, of which the common log is 2.6149, that is, log 412 = 2.6149.

So the mantissa of **log 41.2** is the same as that of the log of 412.

Also, from the highest place value in 41.2, we can get the characteristic of **log 41.2**.

Next, in the last four examples, all the antilogs in the logs are powers.
(Note that a radical can be taken as a power, too. For instance, $\sqrt{2} = 2^{\frac{1}{2}}$, and $\sqrt[3]{5} = 5^{\frac{1}{3}}$.)

And in all the antilogs, the bases share the same sequence of digits, which has 4, 1, and 2 in order. We have: **log 412 = 2.6149**. So the sequence the bases share is the same as the one in 412, of which the log is 2 + 0.6149, where 2 is the characteristic and 0.6149 is the mantissa. So using such information, we can find x in each of the examples.

Besides, in this example, we get to use two log identities below:

$\log_b b = 1$.

$x \log_b A = \log_b A^x$.

In the identities above, A and $b > 0$, but $b \neq 1$, of course.

124

3.0.   log x = 3.6149

We know **log 4.12 = 0.6149**.
The log value, 3.6149 has the same mantissa as that of **log 4.12**, and indicates that the antilog has the same sequence of digits as the one in 4.12, but the highest place value in the antilog is 10^3, which is 1000. So the sequence of digits in x is 4, 1, and 2 in order, and the highest place value in x is 1000. Therefore, x = **4120**.

3.1.   log x = $\bar{1}$.6149

We know **log 4.12 = 0.6149**.
Thus, the log value $\bar{1}$.6149 has the same mantissa as that of **log 4.12**, and thus, indicates that x has the same sequence of digits as the one in 4.12, but the characteristic of the log value is -1, so the highest place value in x is 10^{-1}. Therefore, x = **0.412**.

3.2.   log x = $\bar{2}$.6149

We know **log 4.12 = 0.6149**.
Thus, the log value $\bar{2}$.6149 indicates that x has the same sequence of digits as the one in 4.12, and the highest place value in x is 10^{-2}.

Therefore, x = **0.0412**.

3.3.   log 41.2 = x

We know **log 4.12 = 0.6149**.
To begin with, 41.2 has the same sequence of digits as the one in 4.12.

Thus, **log 41.2** has to have the same mantissa as that of **log 4.12**, which is 0.6149.

Next, the highest place value in 41.2 is 10^1.

So the characteristic of **log 41.2** is 1.
Therefore, x = **1.6149**.

3.4. $\log 0.412^2 = x$

We know **log 4.12 = 0.6149**.

To begin with, 0.412 has the same sequence of digits as the one in 4.12, and the highest place value in it is 10^{-1}.

So the mantissa of the log of 0.412 is 0.6149, and the characteristic of the log is -1.

Thus, we get: **log 0.412 = -1 + 0.6149**.

Next, we have: $P \log_b N = \log_b N^P$.

So we get: $\log 0.412^2 = 2\log 0.412 = 2(-1 + 0.6149) = -2 + 1.2298 = -1 + 0.2298$

$= \overline{1}.2298$, which is x.

3.5. $\log \frac{1}{\sqrt{0.0412}} = x$

We know: **log 4.12 = 0.6149**.

To begin with, 0.0412 has the same sequence of digits as the one in 4.12, and the highest place value in it is 10^{-2}.

So the mantissa of the log of 0.0412 is 0.6149, and the characteristic of the log is -2.

Thus, we get: **log 0.0412 = -2 + 0.6149**.

Next, we have: $P \log_b N = \log_b N^P$. So we get:

$\log \frac{1}{\sqrt{0.0412}} = -\frac{1}{2}\log 0.0412 = -\frac{1}{2}(-2 + 0.6149) = 1 - 0.30745 = 0.69255$, which is x.

3.6.   log 412^x = 1.3542

We know: **log 4.12 = 0.6149.**

To begin with, 412 has the same sequence of digits as the one in 4.12, and the highest place value in it is 10^2.

So the mantissa of the log of 412 is 0.6149, and the characteristic of the log is 2.

Thus, we get: **log 412 = 2 + 0.6149.**

Next, we have: $P \log_b N = \log_b N^P$.

So we get: **log 412^x = x log 412 = x(2.6149) = 1.3542.**

Therefore, $x = \frac{1.3542}{2.6149} = \frac{13542}{26149} \cong$ **0.5179.**

3.7.   log 412^{2x} = $\overline{2}$.3462

We know: **log 4.12 = 0.6149.**

The antilog, 412 has the same sequence of digits as in 4.12, but has the highest place value of 10^2.

So we get: **log 412 = 2.6149.**

Next, we have: $P \log_b N = \log_b N^P$.

Thus, **log 412^{2x} = 2x log 412 = 2x(2.6149) = 5.2258x = $\overline{2}$.3462.**

So we get: **5.2258x = -2 + 0.3462 = -1 – 1 + 0.3462 = -1 – 0.6538 = -1.6538.**

Therefore, $x = \frac{-1.6538}{5.2258} = -\frac{16538}{52258} \cong$ **–0.3165.**

Examples 4 in Common Logs

Note that a small dot·is a multiplication operator unless specified otherwise, e.g. $3 \times 5 = 3 \cdot 5 = 15$.

Note also, that no use of calculator is allowed in this set of examples.

Now, doing the examples below, use **log 2 = 0.3010**, and **log 3 = 0.4771**.

0.   Find the number of digits in 6^{100}.

1.   Find the highest place value in $(\frac{1}{\sqrt{2}})^{25}$.

2.   Find the highest place value and the number in the highest digit in $\frac{27^{100}}{5^{200}}$.

Suggestions or Solutions
To the Problem in the Example 0

Taking the common log of a positive integer, we get a log value, which is composed of two parts. One of the two has to do with the sequence of digits in the integer. Then, what does the other have to do with?

The other is an integer called the characteristic, which can tell us the highest place value in the integer, so what else can the characteristic tell us?

The lowest place value in an integer is 1, and the second lowest place value in it is 10 if it has more than one digit, of course, so the characteristic can tell us the number of all the digits in the integer, too.

Now, assuming **log 2 = 0.3010** and **log 3 = 0.4771**, find the number of digits in 6^{100}.

log 6^{100} = 100 log (2)(3) = 100(log 2 + log 3) = 100(0.3010+0.4771)
= 100(0.7781) = 77.8100.

So the characteristic is 77, and therefore, 6^{100} has 78 digits.

If not quite sure of the idea behind the processes above, follow the steps below:

We know 6^{100} is a positive integer.   And we have: $k \log_b N = \log_b N^k$.

Thus, taking the common log of it, we get **log 6^{100}**, which is **100 log 6**.

So what do we need, now?

We need the value of **log 6**.   How then, can we get it?

We have: $\log 2 = 0.3010$, $\log 3 = 0.4771$, and $\log_b MN = \log_b N + \log_b M$, too.

So we get: **log 6 = log 2 + log 3 = 0.3010 + 0.4771 = 0.7781.**

Thus, we get: **100 log 6 = 100(0.7781) = 77.81 = 77 + 0.8100.**

So the highest place value in 6^{100} is 10^{77}.

The lowest place value in it is 10^0, which is 1.

Therefore, the number of all the digits in it is 78.

In short:

$\log 6^{100}$ = 100 log (2)(3) = 100(log 2 + log 3) = 100(0.3010+0.4771)
= 100(0.7781) = 77.8100.

So the characteristic is 77, and therefore, 6^{100} has 78 digits.

Suggestions or Solutions
To the Problem in the Example 1

Assuming **log 2 = 0.3010** and **log 3 = 0.4771**, find the highest place value in $(\frac{1}{\sqrt{2}})^{25}$.

$\log(\frac{1}{\sqrt{2}})^{25} = 25\log 2^{-\frac{1}{2}} = -\frac{25}{2}\log 2 = -\frac{25}{2}\cdot 0.3010 = -3.7625 = -4 + 0.2375.$

$\Rightarrow$ The highest place value is 10^{-4}.

If not quite sure of the idea behind the processes above, follow the steps below:

Taking the common log of a positive number, we get a log value, which has two parts. Then, of the two parts, what does the highest place value in the number have to do with?

It is the part called the characteristic, which is an integer.
So we want to find the characteristic of the number in question.

Now, the number in question is $(\frac{1}{\sqrt{2}})^{25}$.

So taking the common log of $(\frac{1}{\sqrt{2}})^{25}$, we get: $\log(\frac{1}{\sqrt{2}})^{25} = 25\log 2^{-\frac{1}{2}} = -\frac{25}{2}\log 2$.

We know **log 2 = 0.3010**. So we get:

$-\frac{25}{2}\log 2 = -\frac{25}{2}\cdot 0.3010 = -3.7625 = -3 - 0.7625 = -3 - 1 + 1 - 0.7625$

$= -4 + 0.2375 = \bar{4}.2375.$

Therefore, the highest place value is 10^{-4}, which is one ten thousandth.

In short:

$\log(\frac{1}{\sqrt{2}})^{25} = 25\log 2^{-\frac{1}{2}} = -\frac{25}{2}\log 2 = -\frac{25}{2}\cdot 0.3010 = -3.7625 = -4 + 0.2375.$

$\Rightarrow$ The highest place value is 10^{-4}.

Suggestions or Solutions
To the Problem in the Example 2

Assuming **log 2 = 0.3010** and **log 3 = 0.4771**, find the highest place value and the number in the highest digit in $\frac{27^{100}}{5^{200}}$.

We have **log 2 = 0.3010**, and **log 3 = 0.4771**.

Let $A = \frac{27^{100}}{5^{200}}$. Then, taking **log** of A, we get:

$$\log A = \log \frac{27^{100}}{5^{200}} = \log \frac{3^{300}}{5^{200}} = \log 3^{300} - \log 5^{200} = 300 \log 3 - 200 \log 5.$$

Meanwhile, **log 5** $= \log \frac{10}{2} = $ **log 10 − log 2 = 1 − log 2**.

So we get:

$$\log A = 300 \log 3 - 200(1 - \log 2) = 300 \cdot 0.4771 - 200(1 - 0.3010)$$
$$= 143.13 - 139.8 = 3.3300.$$

Therefore, the highest place value is 10^3, which is 1000.

We have: **log 2 = 0.3010**, and **log 3 = 0.4771**.

So **log 2000 = 3.3010**, and **log 3000 = 3.4771**.

Thus, we get: **log 2000 < log A < log 3000** $\Rightarrow$ **2000 < A < 3000**.

Therefore, the number in the highest digit is 2.

If not quite sure of the idea behind the processes above, follow the steps below:

Taking the common log of a positive number, we get a log value, which has two parts. Then, of the two parts, one is an integer called the characteristic, which can tell us the highest place value in the number.

So to begin with, setting $A = \frac{27^{100}}{5^{200}}$, and taking the common log of A, we get:

$\log \frac{27^{100}}{5^{200}}$, which equals $\log \frac{3^{300}}{5^{200}}$ since $27 = 3^3$.

Next, we have: $\log_b \frac{M}{N} = \log_b M - \log_b N$, and $k \log_b N = \log_b N^k$, too.

So we get: $\log \frac{3^{300}}{5^{200}} = \log 3^{300} - \log 5^{200} = 300 \log 3 - 200 \log 5$.

Meanwhile, $\log 5 = \log \frac{10}{2} = \log 10 - \log 2 = 1 - \log 2$.

Thus, we get: $\log A = 300 \log 3 - 200 (1 - \log 2)$.

We have: $\log 2 = 0.3010$ and $\log 3 = 0.4771$, too.   So we get:

$\log A = 300(0.4771) - 200(1 - 0.3010) = 143.13 - 200(0.699)$
$= 143.13 - 139.8 = 3.3300$.

Therefore, the highest place value in A is 10^3, which means 1000's place.

Next, we want to find the number in the highest digit in A.
That is, what's in the 1000's place?

We have: $\log A = 3.3300$, $\log 2 = 0.3010$, and $\log 3 = 0.4771$.

So we may want to begin with whereabouts the commmon log of A is in a number line.

To begin with, we know the highest place value in both 2000 and 3000 is 3.

So we can see that $\log 2000 = 3.3010$ and that $\log 3000 = 3.4771$.

Thus, since $\log A = 3.3300$, we get:

$$3.3010 < 3.3300 < 3.4771 \Rightarrow \log 2000 < \log A < \log 3000 \Rightarrow 2000 < A < 3000.$$

Therefore, the number in the highest digit is 2. The actual results from a calculator are as follows: $\frac{27^{100}}{5^{200}} = \left(\frac{27}{25}\right)^{100} \cong 2199.7612563$, and $\log 2199.7612563 \cong 3.3424$.

Besides, using a calculator, we can get: $\log A = 3.3300 \Rightarrow A = 10^{3.33} \cong 2137.9620895$.

In short:

We have $\log 2 = 0.3010$, and $\log 3 = 0.4771$.

Let $A = \frac{27^{100}}{5^{200}}$.   Then, taking $\log$ of A, we get:

$$\log A = \log \frac{27^{100}}{5^{200}} = \log \frac{3^{300}}{5^{200}} = \log 3^{300} - \log 5^{200} = 300 \log 3 - 200 \log 5.$$

Meanwhile, $\log 5 = \log \frac{10}{2} = \log 10 - \log 2 = 1 - \log 2$.

So we get:

$$\log A = 300 \log 3 - 200(1 - \log 2) = 300 \cdot 0.4771 - 200(1 - 0.3010)$$
$$= 143.13 - 139.8 = 3.3300.$$

Therefore, the highest place value is 10^3, which is 1000.

We have: $\log 2 = 0.3010$, and $\log 3 = 0.4771$.

So $\log 2000 = 3.3010$, and $\log 3000 = 3.4771$.

Thus, we get: $\log 2000 < \log A < \log 3000 \Rightarrow 2000 < A < 3000$.

Therefore, the number in the highest digit is 2.

B. Number Systems and Logs

Saying just a decimal number, we mean a number indicating a fraction as 0.23.

Saying a decimal number referring to a number system though, we mean a number that is made of powers of 10.

For instance, we can call 256 a decimal number if considering a number system, because we can put it this way: $256 = 2 \cdot 10^2 + 5 \cdot 10^1 + 6 \cdot 10^0$, which is a <u>sum of powers of 10</u>.

Such numbers are said to be in the decimal number system, called the decimal system, for short, and those numbers can be briefly called decimals, too.

And we define a number system using a base.   What then, is the base?

Each digit in a number has to have one single-digit number only. And the base is the number of single-digit integers we can use in each digit in every number in the number system having the base. So for instance, 10 is the base in the decimal number system.

That's because each digit in such a number has one of ten numbers from 0 to 9.
And we can call a decimal number a number to base 10, too.

And if the base is 2, we use in each digit, one of two numbers, which are 0 and 1. And such a number is called a binary number. Also, it can be called a number to base 2.

And from the decimal number's perspective, <u>a binary number is a sum of powers of 2</u>.

For instance, if we put a binary 101 into the decimal system, we get: $1 \cdot 2^2 + 0 \cdot 2^1 + 1 \cdot 2^0$, which is: $2^2 + 0 + 1 = 5$.   So the decimal equivalent of a binary 101 is 5.

And we know taking the common log of a number, we can put the value of the log in terms of the characteristic and mantissa.

So if for instance, a positive integer A has d digits, what is the characteristic c of $\log A$?

It is: $d - 1$. So we can set: $\log A = c + m$, where $c = d - 1$, and $0 \leq m < 1$.

Then, we get: $d - 1 \leq \log A < d$.   So we get: $10^{d-1} \leq A < 10^d$.

Thus, the highest place value in a d-digit integer is: 10^{d-1}, which is 10^c, since $c = d - 1$.

How come can we get this, though: $\log A = c + m$, where $c = d - 1$, and $0 \leq m < 1$?

Let's see now, how we can get this: $\log A = c + m$, where c is an integer, and $0 \leq m < 1$.

If we take a log of a number, the number is positive. Normally, if we take a common log of a number, the number is not only positive but **decimal**, too. In fact, taking a log of a number, we normally use a decimal number weather the log is a common log or not.

Now, taking a common log of a number, we can get the characteristic, which can tell us the highest place value in the number. Such a characteristic can tell us the highest place value in the number because both bases used in the log and the number are 10. The base in a common log is 10, and a decimal number is a number to base 10.

What if however, we want to find the highest place value in a number not decimal?

That is, the number is in a number system to base other than 10.

Then, it depends on the number system the number belongs to.
What do we mean by the number system though?

To begin with, we use a base in a number system as well as in a log and a power.

If a number is decimal, the number is in the decimal number system, where the **base** is **ten**. A decimal number is a number to base ten, and can be briefly called a decimal, too.

In a decimal number, each digit holds one of ten single-digit numbers, and they are 0, 1, 2, 3, 4, 5, 6, 7, 8, and 9.

A binary number is a number to base two, and can be briefly called a binary, too. In a binary, each digit holds one of two single-digit numbers, which are 0 and 1.

Next for instance, in a decimal integer where the base is **10**, the place values are 10^0, 10^1, 10^2, ..., 10^{n-1} if the integer has n digits.

By the same token, in a binary integer where the base is **2**, the place values are 2^0, 2^1, 2^2, ..., 2^{n-1} if the binary integer has n digits.

For more detailed explanations on such a base, refer to **BASES** in **NUMBER SYSETMS**.

Suppose now, we want to find the highest place value in a number that is not decimal. Suppose for instance, we want to find the highest place value in a binary 101.

Then, the highest place value is not 10^2 but 2^2.   How come?

That's because 101 is not $1 \cdot 10^2 + 0 \cdot 10^1 + 1 \cdot 10^0 = 100 + 1$ but $1 \cdot 2^2 + 0 \cdot 2^1 + 1 \cdot 2^0$, which is $4 + 1 = 5$, which is the decimal equivalent of a binary 101.

Now, what if we want to find the highest place value in a binary, which is the binary equivalent of a decimal 20435910?

Then, of course, converting the decimal to the binary equivalent, we can see the highest place value counting the digits in the binary equivalent.

Not all decimals can get readily converted, though.
Converting 20435910 to the binary equivalent, we might not have to spend too much time, but converting a decimal like 3^{3412}, we probably cannot avoid spending a long time if not using a calculator or computer.

We don't actually need to use such a gadget, though.
We just run math, instead. Running it, we can readily get the highest place value taking a log of the decimal 3^{3412} to base 2.   How?

Suppose A is the decimal 3^{3412}, which is the decimal equivalent of the binary in which we want to find the highest place value.

Then, taking a log of A to base 2, we get $\log_2 A$, which can be put the way below:

$\log_2 A = u + v$, where u is an integer, and $0 \leq v < 1$.

Then, the highest place value in the binary equivalent of the decimal 3^{3412} is: 2^u.

More specifically, if the binary equivalent is an n-digit binary, u is: $n - 1$.

That's because the highest place value an n-digit binary is 2^{n-1}.

For instance, in a 3-digit binary 101, the highest place value is: 2^{3-1}, which is: 2^2.

So if A is the decimal equivalent of an n-digit binary, taking a log of A to base 2, we get:

$\log_2 A = u + v$, where $u = n - 1$, and $0 \leq v < 1$.   How come though, it is the case?

Suppose first, A is a k-digit integer positive and decimal, and d_m is each digit in A.

Then, expanding A in terms of digits, we can put it the way below:

$A = d_0 10^0 + d_1 10^1 + d_2 10^2 + \ldots + d_{k-1} 10^{k-1}.$

Normally, putting a number in writing, we just show its digits, so we put A in this form: $d_{k-1}d_{k-2}d_{k-3} \ldots d_2d_1d_0$, of which the expansion is $d_010^0 + d_110^1 + d_210^2 + \ldots + d_{k-1}10^{k-1}$.

For instance, $35028 = 8{\cdot}10^1 + 2{\cdot}10^1 + 0{\cdot}10^2 + 5{\cdot}10^3 + 3{\cdot}10^4$, which is a 5-digit decimal integer.

Suppose next, that we convert the decimal integer A into an n-digit binary integer B, and that b_m is each digit in B.

That is, B is equivalent to A, and can only look different, so both are the same in value.

Then, we can put B in $b_{n-1}b_{n-2}b_{n-3} \ldots b_2b_1b_0$, and call B, the binary equivalent of A.

So we can set: $A = b_02^0 + b_12^1 + b_22^2 + \ldots + b_{n-1}2^{n-1}$, too.

That is because A is the same as B in value. Thus, we can set:

$$A = d_010^0 + d_110^1 + d_210^2 + \ldots + d_{k-1}10^{k-1} = b_02^0 + b_12^1 + b_22^2 + \ldots + b_{n-1}2^{n-1} = B.$$

So as a decimal integer, the highest place value in A is 10^{k-1}.

And as a binary integer, the highest place value in A is 2^{n-1}.

Now, let's quickly go over the ideas above.

Suppose A is a k-digit integer decimal and positive, and d_m is each digit in A. Then, expanding A, we get:

$$A = d_{k-1}d_{k-2}d_{k-3} \ldots d_2d_1d_0 = d_010^0 + d_110^1 + d_210^2 + \ldots + d_{k-1}10^{k-1}.$$

Suppose next, B is an n-digit binary integer, which is the binary equivalent of A, and b_m is each digit in B. Then, expanding B, we get:

$$B = b_{n-1}b_{n-2}b_{n-3} \ldots b_2b_1b_0 = b_02^0 + b_12^1 + b_22^2 + \ldots + b_{n-1}2^{n-1}.$$

Now, A is equivalent to B, and can only look different. So we can set:

A

$= d_{k-1}d_{k-2}d_{k-3} \ldots d_2 d_1 d_0$

$= d_0 10^0 + d_1 10^1 + d_2 10^2 + \ldots + d_{k-1} 10^{k-1}$

$= b_0 2^0 + b_1 2^1 + b_2 2^2 + \ldots + b_{n-1} 2^{n-1}$

$= b_{n-1} b_{n-2} b_{n-3} \ldots b_2 b_1 b_0$

$= B$.

In sum, we have:

$A = d_0 10^0 + d_1 10^1 + d_2 10^2 + \ldots + d_{k-1} 10^{k-1} = b_0 2^0 + b_1 2^1 + b_2 2^2 + \ldots + b_{n-1} 2^{n-1} = B$.

Then first, taking a common log of A, we can get:

$\log_{10} A = \log A = c + m$, where c is an integer, and $0 \leq m < 1$. (We will see why shortly.)

Then anyway, c is the characteristic, and m is the mantissa. More specifically, c is the characteristic for decimal numbers, and m is the mantissa for decimal numbers.

So as a decimal integer, the highest place value in A is 10^c. (We will see why shortly.)
Thus anyway, the characteristic $c = k - 1$.
That is because if A is a k-digit decimal integer, the highest place value in A is 10^{k-1}.

Next, taking a log of A to base 2, we can get:
$\log_2 A = u + v$, where u is an integer, and $0 \leq v < 1$.

Then, u is the characteristic for binary numbers, and v is the mantissa for binaries.

So the highest place value in the binary equivalent of A is 2^u. (We will see why shortly.)

Thus anyway, the characteristic $u = n - 1$.

That is because the highest place value in the n-digit binary equivalent of A is 2^{n-1}.

Now, we are going to see the reason that we can set:

$\log_2 A = u + v$, where u is an integer, and $0 \leq v < 1$.

That is, we will get to see the answers to all the whys above. So to begin with, we know:

$$A = d_0 10^0 + d_1 10^1 + d_2 10^2 + \ldots + d_{k-1} 10^{k-1} = b_0 2^0 + b_1 2^1 + b_2 2^2 + \ldots + b_{n-1} 2^{n-1} = B.$$

So we can set: $A = b_0 2^0 + b_1 2^1 + b_2 2^2 + \ldots + b_{n-1} 2^{n-1}$.   Then first, we can get:

$$A = b_0 2^0 + b_1 2^1 + b_2 2^2 + \ldots + b_{n-1} 2^{n-1} = 2^{n-1} \cdot \frac{b_0 2^0 + b_1 2^1 + b_2 2^2 + \ldots + b_{n-1} 2^{n-1}}{2^{n-1}}.$$

So taking a **log₂** of A, we get:

$$\log_2 A = \log_2 (2^{n-1} \cdot \frac{b_0 2^0 + b_1 2^1 + b_2 2^2 + \ldots + b_{n-1} 2^{n-1}}{2^{n-1}})$$

$$= \log_2 2^{n-1} + \log_2 \frac{b_0 2^0 + b_1 2^1 + b_2 2^2 + \ldots + b_{n-1} 2^{n-1}}{2^{n-1}}$$

$$= (n-1) + \log_2 \frac{b_0 2^0 + b_1 2^1 + b_2 2^2 + \ldots + b_{n-1} 2^{n-1}}{2^{n-1}}.$$

Meanwhile, we have:

$$\frac{b_0 2^0 + b_1 2^1 + b_2 2^2 + \ldots + b_{n-1} 2^{n-1}}{2^{n-1}} = b_0 2^{0-n+1} + b_1 2^{1-n+1} + b_2 2^{2-n+1} + \ldots + b_{n-2} 2^{n-2-n+1} + b_{n-1}$$

$$= b_0 2^{1-n} + b_1 2^{2-n} + b_2 2^{3-n} + \ldots + b_{n-2} 2^{-1} + b_{n-1} = \frac{b_0}{2^{n-1}} + \frac{b_1}{2^{n-2}} + \frac{b_2}{2^{n-3}} + \ldots + \frac{b_{n-2}}{2} + b_{n-1}.$$

Now, for $m = 0, 1, 2, \ldots, n-1$, every b_m is 1 at most, since 1 and 0 are all the digits that can be used in a binary number, and for the same reason, every b_m is 0 at least.

Thus, $\frac{b_0}{2^{n-1}} + \frac{b_1}{2^{n-2}} + \frac{b_2}{2^{n-3}} + \ldots + \frac{b_{n-2}}{2} + b_{n-1}$ is at most $\frac{1}{2^{n-1}} + \frac{1}{2^{n-2}} + \frac{1}{2^{n-3}} + \ldots + \frac{1}{2} + 1$, and is at least $\frac{0}{2^{n-1}} + \frac{0}{2^{n-2}} + \frac{0}{2^{n-3}} + \ldots + \frac{0}{2} + 0 = 0.$

Next, we have: $1 \le \dfrac{1}{2^{n-1}} + \dfrac{1}{2^{n-2}} + \dfrac{1}{2^{n-3}} + ... + \dfrac{1}{2} + 1 < 2$.

In other words, we have: $1 \le \dfrac{1}{2^{n-1}} + ... + \dfrac{1}{2^3} + \dfrac{1}{2^2} + \dfrac{1}{2} + 1 < 2$. How come, though?

We have: $\dfrac{1}{2^{n-1}} + ... + \dfrac{1}{2^2} + \dfrac{1}{2} + 1 = \dfrac{1\{1-(\frac{1}{2})^n\}}{1-\frac{1}{2}} = \dfrac{1-\frac{1}{2^n}}{\frac{1}{2}} = 2(1-\dfrac{1}{2^n}) = 2 - \dfrac{1}{2^{n-1}}$. How come?

Suppose $S = a + ar + ar^2 + ar^3 + ... + ar^{k-2} + ar^{k-1}$.

Then, $rS = ar + ar^2 + ar^3 + ... + ar^{k-1} + ar^k$.

Thus, we get: $S - rS = S(1 - r)$

$= (a + ar + ar^2 + ar^3 + ... + ar^{k-2} + ar^{k-1}) - (ar + ar^2 + ar^3 + ... + ar^{k-1} + ar^k)$

$= a - ar^k = a(1 - r^k)$.

So we get: $S(1 - r) = a(1 - r^k)$.

Thus, we get: $S = a + ar + ar^2 + ar^3 + ... + ar^{k-2} + ar^{k-1} = \dfrac{a(1-r^k)}{1-r}$.

Now, we have: $\dfrac{1}{2^{n-1}} + ... + \dfrac{1}{2^3} + \dfrac{1}{2^2} + \dfrac{1}{2} + 1 = 1 + 1 \cdot \dfrac{1}{2} + 1 \cdot \dfrac{1}{2^2} + 1 \cdot \dfrac{1}{2^3} + ... + 1 \cdot \dfrac{1}{2^{n-1}}$.

So taking 1 as a as 1 and taking $\frac{1}{2}$ as r in the sum S above, we get:

$$S = \dfrac{a(1-r^k)}{1-r} = \dfrac{1\{1-(\frac{1}{2})^n\}}{1-\frac{1}{2}} = \dfrac{1-\frac{1}{2^n}}{\frac{1}{2}} = 2(1-\dfrac{1}{2^n}) = 2 - \dfrac{1}{2^{n-1}}.$$

Next, since n is the number of digits in an integer, n is an integer ≥ 1.

So if $n = 1$, we get: $2 - \dfrac{1}{2^{n-1}} = 2 - \dfrac{1}{2^{1-1}} = 2 - 1 = 1$, and if n is large, we get: $2 - \dfrac{1}{2^{n-1}} \approx 2$.

Thus, we get: $1 \le 2 - \dfrac{1}{2^{n-1}} < 2$, so we get: $1 \le \dfrac{1}{2^{n-1}} + \dfrac{1}{2^{n-2}} + \dfrac{1}{2^{n-3}} + ... + \dfrac{1}{2} + 1 < 2$.

We know: $\dfrac{b_0}{2^{n-1}} + \dfrac{b_1}{2^{n-2}} + \dfrac{b_2}{2^{n-3}} + ... + \dfrac{b_{n-2}}{2} + b_{n-1}$ is at most $\dfrac{1}{2^{n-1}} + \dfrac{1}{2^{n-2}} + \dfrac{1}{2^{n-3}} + ... + \dfrac{1}{2} + 1$.

So we get: $1 \le \dfrac{b_0}{2^{n-1}} + \dfrac{b_1}{2^{n-2}} + \dfrac{b_2}{2^{n-3}} + ... + \dfrac{b_{n-2}}{2} + b_{n-1} < 2$.

We have: $\dfrac{b_0 2^0 + b_1 2^1 + b_2 2^2 + ... + b_{n-1} 2^{n-1}}{2^{n-1}} = \dfrac{b_0}{2^{n-1}} + \dfrac{b_1}{2^{n-2}} + \dfrac{b_2}{2^{n-3}} + ... + \dfrac{b_{n-2}}{2} + b_{n-1}$.

So we get: $1 \le \dfrac{b_0 2^0 + b_1 2^1 + b_2 2^2 + ... + b_{n-1} 2^{n-1}}{2^{n-1}} < 2$.

Thus, we get: $\log_2 1 \le \log_2 \dfrac{b_0 2^0 + b_1 2^1 + b_2 2^2 + ... + b_{n-1} 2^{n-1}}{2^{n-1}} < \log_2 2$.

So we get: $0 \le \log_2 \dfrac{b_0 2^0 + b_1 2^1 + b_2 2^2 + ... + b_{n-1} 2^{n-1}}{2^{n-1}} < 1$.

Thus, we get: $\log_2 A = (n-1) + \log_2 \dfrac{b_0 2^0 + b_1 2^1 + b_2 2^2 + ... + b_{n-1} 2^{n-1}}{2^{n-1}}$.

Now, setting $u = n - 1$, and $v = \log_2 \dfrac{b_0 2^0 + b_1 2^1 + b_2 2^2 + ... + b_{n-1} 2^{n-1}}{2^{n-1}}$, we get:

$\log_2 A = u + v$, where $u = n - 1$, which is an integer, and $0 \le v < 1$.

Thus, we get: $u \le \log_2 A < u + 1 \Rightarrow 2^u \le A < 2^{u+1}$.

Now, we know A is the decimal equivalent of the binary B.
So the highest place value in the binary B is 2^u where u is the characteristic for binary numbers. Of course, u is not a number to base two, that is, not a binary number.

We have: $u = n - 1$, where n is the number of digits in the binary B.

Thus, the highest place value in the binary B is: 2^{n-1}.

And the same is true for $\log_b A$, where A is the decimal equivalent of a number to base b.

That is, we get: $\log_b A = s + t$, where s is an integer, and $0 \leq t < 1$.

Thus, we get: $s \leq \log_b A < s + 1 \Rightarrow b^s \leq A < b^{s+1}$.

So the highest place value in the number to base b which is equivalent to A is: $\underline{b^s}$.

Let's for instance, see if taking a $\log_8$ of 98 works.

$$\log_8 98 = \log_8 64 \cdot \tfrac{98}{64} = \log_8 64 + \log_8 \tfrac{98}{64} = 2 + \log_8 \tfrac{98}{64}.$$

We know: $1 < \tfrac{98}{64} < 8 \Rightarrow \log_8 1 < \log_8 \tfrac{98}{64} < \log_8 8 \Rightarrow 0 < \log_8 \tfrac{98}{64} < 1$.

So we can set: $\log_8 98 = 2 + t$ where $0 < t < 1$, and thus, $s = 2$.

In fact, $98 = 1 \cdot 8^2 + 4 \cdot 8^1 + 2 \cdot 8^0$, which is 142 to base eight.

That is, 98 is the decimal equivalent of an octal 142.

Now, let's see if it is the case where $8^s \leq 98 < 8^{s+1}$.

We know: $s = 2$, so $8^s = 8^2 = 64$, and $8^{s+1} = 8^3 = 512$.   So it is the case.

In general, we can set: $\log_b A = s + t$, where s is an integer, and $0 \leq t < 1$.
Then, the integer s is called the *index*, and t is called the mantissa of $\log_b A$.

More specifically:

The index s can be called the index for the numbers to base b.
The mantissa t can be called the mantissa for the numbers to base b.
So the highest place value in the number to base b is: b^s, where s is the index (called the characteristic, too) for the numbers to base b.

Therefore, $\log_b A = s + t \Leftrightarrow s \leq \log_b A < s + 1 \Leftrightarrow b^s \leq A < b^{s+1}$, where A is positive, of course, and is the decimal equivalent of a number to base b.

Examples 1 in Number Systems and Logs

Note that a small dot·is a multiplication operator unless specified otherwise,
e.g. $3 \times 5 = 3{\cdot}5 = 15$.

Note also, that no use of calculator is allowed in this set of examples.

0.   Assuming: **$\log 2 = 0.3010$**, and **$\log 3 = 0.4771$**, and **2^n** a 30-digit integer, find **n**.

1.   It is known that 47^{100} has 168 digits. Find the number of digits in 47^{17}.

2.   Suppose a is a positive integer, but b is just a positive number. Suppose also, a^{50} has 42 digits, and b^{-50} has 35 zeroes before the first nonzero appears below the decimal point. Find the number of digits above the decimal point in $(ab)^{10}$.

3.   Assuming that $\log x^2$ and $\log \frac{1}{x}$ both have the same mantissa, find the antilog x in $\log x = 2 + m$, where $0 \le m < 1$.

Suggestions or Solutions
To the Problem in the Example 0

Assuming log 2 = 0.3010, log 3 = 0.4771, and 2^n is a 30-digit integer, find n.

2^n is a 30-digit integer $\Rightarrow n$ is an integer, too, and the characteristic $= 30 - 1 = 29$.

So we get: $29 \le \log 2^n < 30$.

And we have: $\log 2^n = n \log 2$, where n is an integer, and $\log 2 = 0.3010$. So we get:

$$29 \le \log 2^n < 30 \Rightarrow 29 \le n \log 2 < 30 \Rightarrow \frac{29}{\log 2} \le n < \frac{30}{\log 2} \Rightarrow 96.3455 < n < 99.6678.$$

Therefore, $n = 97, 98$, or 99.

If not quite sure of the idea behind the processes above, follow the steps below:

To begin with, if 2^n is an integer, n is an integer, too. And next, 2^n is positive, of course.

So next, assuming 2^n is a d-digit integer, we get: $\log 2^n = (d - 1) + m$, where $0 \le m < 1$.

And the characteristic of $\log 2^n$ is: $d - 1$.

So if 2^n has 30 digits, the characteristic of $\log 2^n$ is: $30 - 1 = 29$.

In other words, $29 \le \log 2^n < 30$. (So $10^{29} \le 2^n < 10^{30}$, and thus, 2^n is a 30-digit integer.)

And next, we have: $\log 2^n = n \log 2$, too.

So we get: $29 \le \log 2^n < 30 \Rightarrow 29 \le n \log 2 < 30 \Rightarrow \frac{29}{\log 2} \le n < \frac{30}{\log 2}$.

And we have: $\log 2 = 0.3010$, too.   And also, we know n is an integer.

So we get: $\frac{29}{0.3010} \le n < \frac{30}{0.3010} \Rightarrow 96.3455 < n < 99.6678$, and thus, $n = 97, 98$, or 99.

Suggestions or Solutions
To the Problem in the Example 1

It is known that 47^{100} has 168 digit. Find the number of digits in 47^{17}.

$167 \leq \log 47^{100} < 168 \Rightarrow 1.67 \leq \log 47 < 1.68$

$\Rightarrow 17(1.67) \leq \log 47^{17} < 17(1.68) \Rightarrow 28.3900 \leq \log 47^{17} < 28.56$.

Therefore, 47^{17} has 29 digits.

If not quite sure of the idea behind the processes above, follow the steps below:

Suppose A is the integer. What then, does the number of digits in A have to do with?

It has much to do with the highest place value in A.

The highest place value in an *n*-digit integer is 10^{n-1}.

So assuming that 10^8 is the highest place value in A, we can see A is a 9-digit integer.

Thus, finding the highest place value in A, we can get the number of digits in A.

Taking a log of a number, positive, of course, we can get the information on the highest place value in the number, and if the number is an integer, we can get the information on the number of digits in it, too. The information is called the index or the characteristic.

Now, we want to find the number of digits in 47^{17}, which is an integer decimal.

And we have a log identity, where $x \log A = \log A^x$.

Thus, multiplying **log 47** by 17, we can find the characteristic of **$\log 47^{17}$**.

We need though, to get the value of **log 47**, first.   How then, can we get it?

Actually, we don't have to get the whole thing about **log 47**.

We need only the characteristic of it, since we are after the number of the digits only.

We can find the characteristic by means of the information on another number 47^{100}, and the information is that it has 168 digits. So the characteristic of **log** 47^{100} is 167.

Thus, we can set: **log** 47^{100} = 167 + *m*, where 0 ≤ *m* < 1.

So we get: 167 ≤ **log** 47^{100} < 168 ⇒ 167 ≤ 100 **log** 47 < 168 ⇒ 1.67 ≤ **log** 47 < 1.68.

Thus, we can now get the characteristic of **log** 47^{17}.

And we have a log identity, *k* **log** *A* = **log** A^k.   So **17 log** 47 = **log** 47^{17}.   Thus, we get:

1.67 ≤ log 47 < 1.68 ⇒ 1.67(17) ≤ 17 log 47 < 1.68(17) ⇒ 28.39 ≤ log 47^{17} **< 28.56.**

Thus, we can see that the characteristic of **log** 47^{17} is **28**.

So the highest place value in 47^{17} is 10^{28}, and therefore, 47^{17} has **29** digits.

In short:

167 ≤ log 47^{100} **< 168 ⇒ 1.67 ≤ log 47 < 1.68**

⇒ 17(1.67) ≤ log 47^{17} **< 17(1.68) ⇒ 28.3900 ≤ log** 47^{17} **< 28.56.**

Therefore, 47^{17} has 29 digits.

Suggestions or Solutions
To the Problem in the Example 2

Suppose a is a positive integer, but b is just a positive number.
Suppose also, a^{50} has 42 digits, and b^{-50} has 35 zeroes before the first nonzero appears below the decimal point.

Find the number of digits above the decimal point in $(ab)^{10}$.

$41 \leq \log a^{50} < 42 \Rightarrow \frac{41}{50} \leq \log a < \frac{42}{50} \Rightarrow 0.82 \leq \log a < 0.84$.

$-36 \leq \log b^{-50} < -35 \Rightarrow \frac{36}{50} \geq \log b > \frac{36}{50} \Rightarrow 0.7 < \log b \leq 0.72$.

Thus, $1.52 < \log a + \log b < 1.56 \Rightarrow 15.2 \leq \log (ab)^{10} < 15.6 \Rightarrow$ 16 digits.

If not quite sure of the idea behind the processes above, follow the steps below:

The number of digits in a number has to do with the highest place value in it.

If the number is decimal, the characteristic in the common log of the number can give us the highest place value.

Now, we want to find the number of digits above the decimal point in a number $(ab)^{10}$.

To begin with, finding **log ab**, first, we can readily get the characteristic of **log $(ab)^{10}$**.

We have information on a^{50} and b^{-50}.

The first of the two is an integer, and has 42 digits, so the highest place value in it is 10^{41}.

And the other is just a number, and has 35 zeros before the first nonzero appears below the decimal point, so the highest place value in it is 10^{-36}.

In other words, the charateristic of $\log a^{50}$ is **41**, and that of $\log b^{-50}$ is **-36**.

Let's now, begin with $\log a^{50}$. Then, we can get:

$41 \leq \log a^{50} < 42 \Rightarrow 41 \leq 50 \log a < 42 \Rightarrow \frac{41}{50} \leq \log a < \frac{42}{50} \Rightarrow 0.82 \leq \log a < 0.84.$

Let's next, move on to $\log b^{-50}$, where the characteristic is -36.

Note first, that the mantissa of any log is supposed to be *nonnegative*, and it is.
More specifically, if m is a mantissa, we have: $0 \leq m < 1$. In this case therefore:

We need to have: **-36 $\leq \log b^{-50} <$ -35** and **not** $-37 \leq \log b^{-50} < -36$.

That's because $\log b^{-50} = -36 + m$, where $0 \leq m < 1$. So we get:

$-36 \leq \log b^{-50} < -35 \Rightarrow -36 \leq (-50) \log b < -35 \Rightarrow \frac{36}{50} \geq \log b > \frac{36}{50} \Rightarrow 0.7 < \log b \leq 0.72.$

Now, we have: $0.82 \leq \log a < 0.84$, and $0.7 < \log b \leq 0.72$. So we get:

$1.52 < \log a + \log b < 1.56 \Rightarrow 1.52 < \log ab < 1.56 \Rightarrow (1.52)10 \leq 10 \log ab < 10(1.56)$

$\Rightarrow 15.2 \leq \log (ab)^{10} < 15.6.$

Thus, the characteristic is 15, so we can see that the highest place value in $(ab)^{10}$ is 10^{15}.

Therefore, it has 16 digits above the decimal point.

In short:

$41 \leq \log a^{50} < 42 \Rightarrow \frac{41}{50} \leq \log a < \frac{42}{50} \Rightarrow 0.82 \leq \log a < 0.84.$

$-36 \leq \log b^{-50} < -35 \Rightarrow \frac{36}{50} \geq \log b > \frac{36}{50} \Rightarrow 0.7 < \log b \leq 0.72.$

Thus, $1.52 < \log a + \log b < 1.56 \Rightarrow 15.2 \leq \log (ab)^{10} < 15.6 \Rightarrow$ 16 digits.

Suggestions or Solutions
To the Problem in the Example 3

Assuming that $\log x^2$ and $\log \frac{1}{x}$ both have the same mantissa, find the antilog x in $\log x = 2 + m$, where $0 \leq m < 1$.

$\log x = 2 + m$, and $0 \leq m < 1 \Rightarrow$ **(1)** $\log x^2 = 4 + 2m$.

$\log x = 2 + m$, and $0 \leq m < 1 \Rightarrow$ **(2)** $\log \frac{1}{x} = -2 - m = -2 - 1 + 1 - m = -3 + (1 - m)$.

Suppose first, $m = 0$. Then, $\log x = 2 \Rightarrow x = 100$.

Suppose next, $m \neq 0$. Then:

For $\log x^2$, we get: $0 < 2m < 1 \Rightarrow 0 < m < \frac{1}{2}$.

For $\log \frac{1}{x}$, we get: $0 < 1 - m < 1 \Rightarrow -1 < -m < 0 \Rightarrow 0 < m < 1$.

Thus, (both $0 < m < \frac{1}{2}$ and $0 < m < 1$) $\Rightarrow 0 < m < \frac{1}{2}$.

Let's now, begin with the case where $0 < m < \frac{1}{2}$.   Then:

From **(1)**, we get: $\log x^2 = 4 + 2m$, and from **(2)**, we get: $\log \frac{1}{x} = -3 + (1 - m)$.

Then, $2m = 1 - m \Rightarrow 3m = 1 \Rightarrow m = \frac{1}{3} \Rightarrow \log x = 2 + \frac{1}{3} = \frac{7}{3} \Rightarrow x = 10^{\frac{7}{3}}$.

Let's next, move on to the case where $\frac{1}{2} \leq m < 1$.

Then first, from **(1)**, we get: $1 \leq 2m < 2 \Rightarrow 1 \leq$ **mantissa** < 2, which is not allowed.

So we cannot take $2m$ for the mantissa of $\log x^2$, and we need to get the proper one.

From **(1)**, $\log x^2 = 4 + 2m = 4 + 1 - 1 + 2m = 5 + (2m - 1)$.

Let's see if $2m - 1$ can be a mantissa.

$\frac{1}{2} \leq m < 1 \Rightarrow 1 \leq 2m < 2 \Rightarrow 0 \leq 2m - 1 < 1$, which is allowed, since $0 \leq$ **mantissa** < 1.

Next, we have: $\log \frac{1}{x} = -3 + (1 - m)$.   Let's see if $(1 - m)$ can be a mantissa.

$\frac{1}{2} \le m < 1 \Rightarrow -1 < -m \le -\frac{1}{2} \Rightarrow 0 < 1 - m \le \frac{1}{2}$, which is allowed, since $0 \le \textbf{mantissa} < 1$.

Then, $2m - 1 = 1 - m \Rightarrow 3m = 2 \Rightarrow m = \frac{2}{3} \Rightarrow \log x = 2 + m = 2 + \frac{2}{3} = \frac{8}{3} \Rightarrow x = 10^{\frac{8}{3}}$.

Therefore, $x = 100$, $10^{\frac{7}{3}}$, or $10^{\frac{8}{3}}$.

If not quite sure of the idea behind the processes above, follow the steps below:

The equation above is for m rather than x, so it is for the mantissa rather than the antilog. That is, finding the mantissa m, we can readily get the value of the antilog x.

Now, we have a condition that mantissas in both $\log x^2$ and $\log \frac{1}{x}$ are the same.

So first, we may want to extract the mantissas from both $\log x^2$ and $\log \frac{1}{x}$.

We can do so putting $\log x^2$ and $\log \frac{1}{x}$ in terms of the characteristic and mantissa.

Let's now, begin with $\log x^2$.

We have: $\log x = 2 + m$, where $0 \le m < 1$, and we have: $\log x^2 = 2 \log x$, too.

So we can get: $\log x = 2 + m \Rightarrow 2 \log x = \log x^2 = 2(2 + m) = 4 + 2m \Rightarrow \log x^2 = 4 + 2m$.

Next, we have: $\log \frac{1}{x} = \log x^{-1} = (-1) \log x = -\log x$. So we get:

$\log x = 2 + m \Rightarrow -\log x = -(2 + m) = -2 - m = -2 - 1 + 1 - m = -3 + (1 - m)$. Why?

A mantissa cannot be negative, so the mantissa of $-\log x$ is not $-m$ but $1 - m$ because we have: $0 \le m < 1$.

Still not quite sure why the mantissa of $\log \frac{1}{x}$, which is $-\log x$, is not $-m$ but $(1 - m)$?

The bottom line is: $0 \le$ **mantissa** < 1, so basically, a mantissa cannot be negative.

Thus, no mantissa is negative no matter what log it may belongs to.
So first, we have: $0 \le m < 1$, where m is the mantissa of **log** x, which is a log.

Next, since we have: $0 \le m < 1$, we get: $-1 < -m \le 0$, so $-m$ is negative or 0, and therefore, is not eligible for a mantissa of any log unless it is 0.

Now, we have: $-\mathbf{log}\, x = -2 - m$, where $0 \le m < 1$, and $-\mathbf{log}\, x$ is a log.

So $-m$ cannot be the mantissa of $-\mathbf{log}\, x$ because $-m$ is negative or 0 since $-1 < -m \le 0$.

To keep the mantissa nonnegative, we can put the above in such a way as follows:

$-\mathbf{log}\, x = -3 + (1 - m)$, which is equivalent to $-2 - m$.

Then, we can see that the mantissa is $1 - m$ because $0 < 1 - m \le 1$.
What about the case where $1 - m = 1$, though?

It will be the case where $m = 0$, and we can take care of it separately.

So before we move on, we may want to set up two different cases for m.

One is that $m = 0$, and the other is that $m \ne 0$.

First, if $m = 0$, then $\mathbf{log}\, x = 2 + m = -2$, so we can simply get: $x = 10^2$ by the definition for logs. So $x = 10^2$ is a solution.

Suppose next, that $m \ne 0$, that is, we have: $0 < m < 1$, and not $0 \le m < 1$.

Then first, the mantissa of $\mathbf{log}\frac{1}{x} = -\mathbf{log}\, x$ is not $-m$ but $(1 - m)$.

Next, in $\mathbf{log}\, x^2 = 4 + 2m$, the mantissa is $2m$, which however, looks problematic.

Since $m \ne 0$, we have $0 < m < 1$. So, we get:

$0 < m < 1 \Rightarrow 0 < 2m < 2$, which is **not** OK, because we need to have: $0 \leq$ **mantissa** < 1.

Then, we can do this: $0 \leq$ **mantissa** $< 1 \Rightarrow 0 \leq 2m < 1 \Rightarrow 0 \leq m < \frac{1}{2}$.

So we can take **2m** for the mantissa of $\log x^2$ in the case where $0 < m < \frac{1}{2}$. Why is 0 not included in the interval?

We've already taken care of the case where $m = 0$, and $x = 10^2$ is the solution in the case. So we still need to take care of the case where $0 < m < 1$.

Thus, we may want to set up again, two different cases for **m**.

One is that $0 < m < \frac{1}{2}$, and the other is that $\frac{1}{2} \leq m < 1$.

Let's however, check to see if $(0 < m < \frac{1}{2})$ satisfies the mantissa of $\log \frac{1}{x}$, too.

That's because the mantissas of both $\log x^2$ and $\log \frac{1}{x}$ are the same.

We know that the mantissa of $\log \frac{1}{x}$ is $1 - m$. So we want to see if $1 - m$ is OK.

$0 < m < \frac{1}{2} \Rightarrow -\frac{1}{2} < -m < 0 \Rightarrow \frac{1}{2} < 1 - m < 1$, which is OK, since $0 <$ **mantissa** < 1.

Let's now, begin with the case where $0 < m < \frac{1}{2}$.

In $\log x^2$, the mantissa is **2m**, and in $\log \frac{1}{x}$, the mantissa is $(1 - m)$.

The two mantissas are the same. So we get: $2m = 1 - m \Rightarrow 3m = 1 \Rightarrow m = \frac{1}{3}$.

We have: $\log x = 2 + m$. So we get: $\log x = 2 + \frac{1}{3} = \frac{7}{3} \Rightarrow x = 10^{\frac{7}{3}}$ by the definition for logs.

Let's next, move on to the case where $\frac{1}{2} \leq m < 1$.

Taking care of **log x^2** first, we need to find the proper mantissa of it because **$2m$** cannot be the mantissa in this case. So finding the proper one, we can put the log this way, too:

$$\log x^2 = 4 + 2m = 4 + 1 - 1 + 2m = 5 + (2m - 1).$$

Next, checking to see if **$(2m - 1)$** can be the proper mantissa, we get:

$\frac{1}{2} \le m < 1 \Rightarrow 1 \le 2m < 2 \Rightarrow 0 \le 2m - 1 < 1$, which is OK, since **$0 \le$ mantissa < 1**.

Next, we have: **$\log \frac{1}{x} = -3 + (1 - m)$**.

So checking to see if **$(1 - m)$** can be a proper mantissa, we get:

$\frac{1}{2} \le m < 1 \Rightarrow -1 < -m \le -\frac{1}{2} \Rightarrow 0 < 1 - m \le \frac{1}{2}$, which is OK, since **$0 \le$ mantissa < 1**.

Since the two mantissas are the same, we get: **$2m - 1 = 1 - m \Rightarrow 3m = 2 \Rightarrow m = \frac{2}{3}$**.

And we have **$\log x = 2 + m$**.

So **$\log x = 2 + \frac{2}{3} = \frac{8}{3} \Rightarrow x = 10^{\frac{8}{3}}$** by the definition for logs.

Thus, **$x = 100, 10^{\frac{7}{3}}$, or $10^{\frac{8}{3}}$**. For reference, **$10^{\frac{7}{3}} \cong 215.443469$, and $10^{\frac{8}{3}} \cong 464.158883$**.

In short:

$\log x = 2 + m$, and **$0 \le m < 1 \Rightarrow$ (1) $\log x^2 = 4 + 2m$**.
$\log x = 2 + m$, and **$0 \le m < 1 \Rightarrow$ (2) $\log \frac{1}{x} = -2 - m = -2 - 1 + 1 - m = -3 + (1 - m)$**.

Suppose first, **$m = 0$**. Then, **$\log x = 2 \Rightarrow x = 100$**.
Suppose next, **$m \ne 0$**. Then:

For $\log x^2$, we get: $0 < 2m < 1 \Rightarrow 0 < m < \frac{1}{2}$.

For $\log \frac{1}{x}$, we get: $0 < 1 - m < 1 \Rightarrow -1 < -m < 0 \Rightarrow 0 < m < 1$.

Thus, (both $0 < m < \frac{1}{2}$ and $0 < m < 1$) $\Rightarrow 0 < m < \frac{1}{2}$.

Let's now, begin with the case where $0 < m < \frac{1}{2}$. Then:

From **(1)** above, we get: $\log x^2 = 4 + 2m$, and from **(2)**, we get: $\log \frac{1}{x} = -3 + (1 - m)$.

Then, $2m = 1 - m \Rightarrow 3m = 1 \Rightarrow m = \frac{1}{3} \Rightarrow \log x = 2 + \frac{1}{3} = \frac{7}{3} \Rightarrow x = 10^{\frac{7}{3}}$.

Let's next, move on to the case where $\frac{1}{2} \leq m < 1$.

Then first, from **(1)**, we get: $1 \leq 2m < 2 \Rightarrow 1 \leq$ **mantissa** < 2, which is not allowed.

So we cannot take $2m$ for the mantissa of $\log x^2$, and we need to get the proper one.

From **(1)**, $\log x^2 = 4 + 2m = 4 + 1 - 1 + 2m = 5 + (2m - 1)$.

Let's see if $2m - 1$ can be a mantissa.

$\frac{1}{2} \leq m < 1 \Rightarrow 1 \leq 2m < 2 \Rightarrow 0 \leq 2m - 1 < 1$, which is allowed, since $0 \leq$ **mantissa** < 1.

Next, we have: $\log \frac{1}{x} = -3 + (1 - m)$. Let's see if $(1 - m)$ can be a mantissa.

$\frac{1}{2} \leq m < 1 \Rightarrow -1 < -m \leq -\frac{1}{2} \Rightarrow 0 < 1 - m \leq \frac{1}{2}$, which is allowed, since $0 \leq$ **mantissa** < 1.

Then, $2m - 1 = 1 - m \Rightarrow 3m = 2 \Rightarrow m = \frac{2}{3} \Rightarrow \log x = 2 + m = 2 + \frac{2}{3} = \frac{8}{3} \Rightarrow x = 10^{\frac{8}{3}}$.

Therefore, $x = 100$, $10^{\frac{7}{3}}$, or $10^{\frac{8}{3}}$.

Examples 2 in Number Systems and Logs

Note that a small dot · is a multiplication operator unless specified otherwise, e.g. $3 \times 5 = 3 \cdot 5 = 15$.

Note also, that no use of calculator is allowed in this set of examples. Instead, you can use a tool as follows:

$$\log_b A = s + t \Leftrightarrow s \leq \log_b A < s + 1 \Leftrightarrow b^s \leq A < b^{s+1}, \text{ where } 0 \leq t < 1.$$

In the tool above, $(s + 1)$ is the number of digits in a number to base b, and A is positive, of course, and is the decimal equivalent of the number to base b.

Now, assuming: **log 2 = 0.3010**, and **log 3 = 0.4771**, do the examples below:

0. Find the highest place value in the binary equivalent of a decimal integer 3^{21}.

1. Find the highest place value in the decimal equivalent of a 30-digit binary integer.

Suggestions or Solutions
To the Problem in the Example 0

Find the highest place value in the binary equivalent of a decimal integer 3^{21}.
Use log 2 = 0.3010, and log 3 = 0.4771.

$$\log_2 3^{21} = 21 \log_2 3 = 21 \cdot \frac{\log 3}{\log 2} = 21 \cdot \frac{0.4771}{0.3010} \cong 33.2860.$$

Therefore, the highest place value is 2^{33}.

If not quite sure of the idea behind the processes above, follow the steps below:

Converting, of course, the decimal integer 3^{21} into its binary equivalent, then finding the number of digits in the equivalent, we can get the highest place value in the equivalent.

Prior to the conversion though, we need to expand (find the value of) the integer 3^{21}, yet the integer expanded is not a number we can convert into the binary equivalent in a reasonable amount of time.

Expanding it, in fact, we get 10460353203, which is obtained by a calculator, of course. That is, using a calculator, we can get: $10460353203 = 3^{21}$.

We don't have to however, get the expansion, that is, we don't need to get 3^{21} expanded. So we don't need a calculator. And we don't need to convert the expansion, either.

That is to say that we can get the number of digits in the binary equivalent with no use of a calculator.   We have a tool to work with, and the tool is called a logarithm.
How then, can we use such a tool?

Suppose A is the decimal 3^{21}, that is, the decimal equivalent of the binary in which we want to find the highest place value.   Then, taking the log of A to base 2, we can get:

$\log_2 A = s + t$, where s is an integer, and $0 \le t < 1$.

In the expression above, s can be called the index for a binary, and t can be called the mantissa for a binary. Then, the highest place value in the binary equivalent of A is: 2^s.

That is, the highest place value in the binary equivalent of the decimal 3^{21} is: 2^s.

More specifically, if the binary equivalent is an n-digit binary, s is: $n - 1$.

That's because the highest place value an n-digit binary is: 2^{n-1}.

Now, taking a log of 3^{21} to base 2, we get: $\log_2 3^{21} = 21 \log_2 3$.

So putting into the expression above the value of $\log_2 3$, we can get the value of $\log_2 3^{21}$, and in turn, the index. And then, we can get the solution.

How then, can we get the value of $\log_2 3$?

We can take advantage of such an identity as follows: $\log_b A = \dfrac{\log_c A}{\log_c b}$.

Taking advantage of it, we can have: $\log_2 3 = \frac{\log 3}{\log 2}$. Thus, we get:

$\log_2 3^{21} = 21 \log_2 3 = 21 \cdot \frac{\log 3}{\log 2} = 21 \cdot \frac{0.4771}{0.3010} \cong 33.2860$, where 33 is the index for the base two, and 0.2860 is the mantissa for the base two.

Therefore, the highest place value in the binary equivalent of 3^{21} is 2^{33}.

In other words, the binary equivalent has 34 digits.

In short:

$\log_2 3^{21} = 21 \log_2 3 = 21 \cdot \frac{\log 3}{\log 2} = 21 \cdot \frac{0.4771}{0.3010} \cong 33.2860$.

Therefore, the highest place value is 2^{33}.

Suggestions or Solutions
To the Problem in the Example 1

Find the highest place value in the decimal equivalent of a 30-digit binary integer. Use log 2 = 0.3010 and log 3 = 0.4771.

Suppose that A is the decimal equivalent. Then:

$29 \leq \log_2 A < 30 \Rightarrow 2^{29} \leq A < 2^{30} \Rightarrow \log 2^{29} \leq \log A < \log 2^{30}$

$\Rightarrow 29 \log 2 \leq \log A < 30 \log 2 \Rightarrow 29(0.3010) \leq \log A < 30(0.3010)$

$\Rightarrow 8.7290 \leq \log A < 9.0300$

$\Rightarrow$ The characteristic of A is **8** or **9**.

Therefore, the highest place value in the decimal equivalent is 10^8 or 10^9.

If not quite sure of the idea behind the processes above, follow the steps below:

To begin with, what are logarithms about?

They are about exponents. Not only that though, but also it can give us a way we can take looking for the highest place value in a number to any base. So the number can be decimal, binary, octal, etc.

Next, the highest place value in an *n*-digit integer decimal is 10^{n-1}.

So for instance, the highest place value in a 2-digit decimal 89 is: $10^{2-1} = 10^1 = 10$, and $10^{3-1} = 10^2 = 100$ is the highest place value in a 3-digit decimal 903.

What then, is the highest place value in an *n*-digit integer to base *b*?

It is b^{n-1}. For instance, the highest place value in an n-digit integer octal is 8^{n-1}. So for instance, $8^{2-1} = 8^1 = 8$ is the highest place value in a 2-digit octal 37, which is: $3 \cdot 8^1 + 7 \cdot 8^0$, and $2^{6-1} = 2^5 = 32$ is the highest place value in a 6-digit binary 110101, which is: $1 \cdot 2^5 + 1 \cdot 2^4 + 0 \cdot 2^3 + 1 \cdot 2^2 + 0 \cdot 2^1 + 1 \cdot 2^0$.

Next, assuming A is a decimal positive, and taking a $\log_2$ of A, what do we get?

We can get: $\log_2 A = s + t$, where s is an integer can be called the index for the base two, t can be called the mantissa for the base two, and $0 \leq t < 1$.

Then, the highest place value in the binary equivalent of A is 2^s.
That is, the highest place value in the binary equivalent of the decimal 3^{21} is 2^s.

More specifically, if the binary equivalent is an n-digit binary, s is $n - 1$.

That's because the highest place value an n-digit binary is 2^{n-1}.

Suppose now, B is the 30-digit integer binary, and A is the decimal equivalent of B.

Then, the highest place value in B is 2^{29}, so the index of $\log_2 A$ is $30 - 1 = 29$.

Thus, we can set: $\log_2 A = 29 + t$, where t is the mantissa for the base two, and $0 \leq t < 1$.

So we get: $29 \leq \log_2 A < 30$. In other words, we get: $2^{29} \leq A < 2^{30}$.

What we are after solving this problem is however, the highest place value in A, which is a decimal, so the inequality $2^{29} \leq A < 2^{30}$ doesn't give us much help unless we expand 2^{29} and 2^{30}. Such expansions will probably take all day without a calculator.

So it looks like we get to solve for A the inequality: $29 \leq \log_2 A < 30$.

Then, we want to extract **log 2** from $\log_2 A$.   How?

We have such an identity as follows: $\log_x Y = \dfrac{\log_z Y}{\log_z x}$.

So we can set: $\log A = \dfrac{\log A}{\log 2}$, where all the logs are common logs where the base is 10.

Then, we get: $29 \le \log_2 A < 30 \Rightarrow 29 \le \dfrac{\log A}{\log 2} < 30 \Rightarrow 29 \log 2 \le \log A < 30 \log 2$.

And we have: $\log 2 = 0.3010$, too.   So?

So we get: $29 \log 2 \le \log A < 30 \log 2 \Rightarrow 29(0.3010) \le \log A < 30(0.3010)$

$\Rightarrow 8.7290 \le \log A < 9.0300$.

We know we can set: $\log A = s + t$, where s is an integer called the index (characteristic), and $0 \le t < 1$.

So we can set: $8.7290 \le s + t < 9.0300$, where s is an integer, and $0 \le t < 1$.

Thus, we get: $s = 8$ or 9.   And we know the highest place value in A is 10^s.

So the highest place value of A is 10^8 or 10^9.

In other words, the decimal integer A has 9 digits or 10 digits.

In short:

Suppose that A is the decimal equivalent. Then:

$29 \le \log_2 A < 30 \Rightarrow 2^{29} \le A < 2^{30} \Rightarrow \log 2^{29} \le \log A < \log 2^{30}$

$\Rightarrow 29 \log 2 \le \log A < 30 \log 2 \Rightarrow 29(0.3010) \le \log A < 30(0.3010)$

$\Rightarrow 8.7290 \le \log A < 9.0300 \Rightarrow$ The characteristic of A is 8 or 9.

Therefore, the highest place value in the decimal equivalent is 10^8 or 10^9.